Birds of Fray - Top Gun: Maverick - Special Edition

Rajat Narang

Published by Rajat Narang, 2022.

BIRDS OF FRAY - TOP GUN: MAVERICK - SPECIAL EDITION

First edition. September 1, 2022.

ISBN: 979-8201901394

Written by Rajat Narang.

Also by Rajat Narang

Airbus vs. Boeing: Strategy Wars, Tactical Dogfights, High-G Maneuvers and the Photo Finishes
Birds of Fray - Top Gun: Maverick - Special Edition
Birds of Fray - World's Top 4.5 & 5th Gen Fighter Jet Aircraft Programs

Top Gun: Maverick - Birds of Fray

World's Top 4.5 & 5th Generation Fighter Jet Aircraft Programs –

Handbook with Comparative Analysis

@9000RPM Publishing Works

RAJAT NARANG

Dedicated to:

Top Gun and Maverick & Goose for making Military Aviation Inspirational, Aspirational, Unforgettable and almost Immortal for Generations, including, those having already served, currently serving and the ones yet to Come...

Disclaimer

The Author does not have any kind of financial investments, direct business associations or financial stakes of any sort in any of the fighter jet aircraft programs or companies included & mentioned in this work. It does not promote or endorse any particular company or industry player & its products over others in any manner whatsoever. The overall scope & directional reference compass of this work primarily is the global fighter jet aircraft market with focus on 4.5 and 5th generation aircraft programs.

The analysis has been purely anchored on informational evidences available in the public domain based information sources and utmost effort has been made to ensure fair play to present the most objective view & analysis against the backdrop & context in light of the available facts rather than taking a biased view skewed towards any of the industry players covered.

The opinions expressed throughout are purely author's personal and the views & judgments presented are neither directed or targeted at anyone nor meant as pontification of any sort and are solely based on objective assessment. The names of products, systems and/or brand names mentioned wherever through the work are intellectual property of their respective owners and their mention has been done only & purely for information & creative purposes and it does not indicate or reflect and should not be construed as any kind of promotion or endorsement of any kind. It is also not intended as a technology/technical guide with the inclusion of some technical aspects & performance parameters for analysis done purely from a relative comparative perspective.

The information and facts contained herein are believed to be correct at the time of publication but cannot be guaranteed. All the information presented has been derived from reliable sources, reasonably verified & has been presented purely & solely for informational purposes only. The views expressed throughout are based on broad analysis & assessments only and thus should not be substituted for professional advice & opinion of any kind prior to decision-making. The author expressly disclaims any and all liability to any person or entity pertaining to potential outcomes or consequences of any decisions or actions taken based on the contents of this publication.

Bibliography

Introduction

Fighter jets have remained the traditional backbone & core pivot of the fighting capabilities of Nations & Air Forces globally with their unmatched capabilities for conducting a broad range of critical missions ranging from aerial combat & interdiction, suppression of enemy air defenses (SEAD), ground attack and air dominance/superiority with the doctrine & tactical playbook of aerial warfare having shifted effectively from the erstwhile Dogfighting era to Beyond-Visual-Range engagements now propelled by long range stand-off weapons.

The surge in global defense spending over the recent years, triggered by a rapid deterioration of traditional, rule-based world order and further exacerbated by the Russia-Ukraine war has driven a rapid recapitalization of legacy fighter jet aircraft with modern & latest programs as a key priority. It has led to initiation of the development of a bevy of next generation aircraft programs & fast-tracked procurement of current generation jets across most parts of the globe for expansion of military capabilities & rapid build-up of capacity. The F-35 Lightning II JSF program is witnessing further diversification of its customer base with Switzerland, Finland, Canada and now Germany going for the world's only in-production & in-service 5th generation fighter jet program.

In term of capabilities, the 5th generation aircraft programs are already proving to be clear favorites to replace the legacy platforms with Lockheed Martin's F-35 Lightning II stealth fighter jet program spearheading the charge & the generational shift towards stealth. A lot of action is already underway on the technological front with the 6th generation aircraft programs clearly on the anvil, geared towards the future, with their potential EIS' slated for late 2030s. Continental Europe is bracing for & committing itself to the development of two such 6th gen programs, namely, the Franco-German FCAS program and the Tempest program from the U.K.-Italy-Sweden triad. In the U.S., the NGAD program collectively represents the efforts being made towards the development of 6th generation combat aircraft to be powered by the upcoming Adaptive Cycle Engine (ACE) program featuring the third air stream.

This work is intended to be a quick yet comprehensive resource on the World's Top Fighter Jet Aircraft Programs with focus on a blend of quantitative & qualitative analysis with a comparative, head-to-head comparison on technical specifications & capabilities, in almost virtual dogfights, for respective aircraft classes (heavy. medium & light) and programs followed by comparative analysis of Strengths & Weaknesses based on technical & business perspective, unlike, the usual, traditional & run-of-the mill approaches.

Most of the fighter jet aircraft programs covered in this work have been in-production programs with the exception of the Skunk Works' magnum opus F-22 Raptor program. Additionally, the programs included have been segregated into two classes, medium and heavy fighter jets based on their MTOW. For the purpose of classification, aircraft programs with MTOW between 15,000 lbs to 30,000 lbs have been categorized into Medium fighter jets class while programs with MTOW of over 30,000 lbs have been classified as Heavy jets.

An individual chapter has been dedicated to each aircraft program covering the backdrop & origin, program outline, technical specifications, variants, key operators and latest contract awards etc. The comparative analysis has been conducted based on class with focus on covering technical as well as business aspects of the respective programs for a holistic perspective.

Being the Top Gun: Maverick Special Edition, the spotlight has been on the F-14 Tomcat which has been Maverick's favorite steed for decades which has enabled him to do mind boggling feats, like capturing a MiG from an inverted, upside-down position, and do what he does best in the air, i.e., as Iceman put it aptly & succinctly, 'being dangerous' and 'buzz the tower repeatedly' on his landing approaches despite warnings. A comparison among F-14 Tomcat and modern generation U.S. comparable twin-engine, twin tail fighters on airframe and core parameters is also included (excluding technologies and weapon systems for generational relevance), with the list including, the F-15E Strike Eagle, F/A-18E/F Super Hornet and the F-22 Raptor.

Lastly, domain leadership needs a constant work in progress approach with its conception as a state of static perfection merely being a conceptual fallacy.

Thus, nobody can claim bragging rights of being a domain expert and there is always a scope of improvement & betterment.

Let's Gear up for the Take-Off Roll!

1. Grumman F-14 Tomcat

Image Credits: SSGT Michael D. Gaddis, USAF – ID DF-SD-03-06180 – Public Domain

Description & Primary Role

The F-14 Tomcat program was the U.S. Navy's primary multi-role carrier based twin-engine, twin-tail supersonic fighter jet aircraft program featuring the characteristic technological feature of the 1970s, the variable sweep wing design. The F-14 was named Tomcat in honor of Admiral Thomas F. Connolly, who played a pivotal role in the development of the F-14 as the program head with the Tomcat translating into Tom's Cat referring to Connolly's involvement. Additionally, the name Tomcat was aligned with Grumman's tradition & nomenclature of naming its aircrafts after feline names which included the Wildcat, Hellcat, Tigercat and Bearcat etc.

Origin, Timeline and Scope

The F-14 had its genesis in the Navy's search for a replacement for its primary, carrier-based naval strike fighter of the time, the McDonnell Douglas F-4 Phantom II with the Navy looking for a longer range, high endurance aircraft platform featuring a more powerful radar and long range missiles in order to be capable of intercepting anti-ship missiles carried by Russian bombers, fighters & Subs. The Navy had initiated its Naval Fighter Experimental (VFX) program in the 1960s after the cancellation of the General Dynamics-Grumman Corporation bid with their F-111B bid owing to capability limitations.

The F-14 Tomcat had its maiden flight in the late 1970 and was operationally deployed across Navy carriers in 1974 and carried out aerial superiority, interception and tactical aerial reconnaissance missions successfully along with ground strike missions (following the addition of LANTIRN pods in the 1990s) through the turn of the century till the F/A-18E/F Super Hornet succeeded it in 1999.

Quantities & Units Procured

Around 712 F-14 Tomcats were built and delivered by Grumman Corporation at its production facility based at Calverton, Long Island NY till the production was ultimately wound up in 1991.

Variants and Versions Built

The F-14 Tomcat was developed and built in 3 variants, namely:-

F-14A: The F-14A was the twin-seat interceptor variant developed & delivered to the U.S. Navy with around 478 F-14As delivered to the U.S. Navy while 79 were exported to Iran. The F-14As were later equipped with precision strike munitions.

F-14B: The initial F-14As were subsequently upgraded to the A+ variant with the incorporation of GE's F110-GE-400 engine from 1979 onwards which increased the overall uninstalled thrust by 40% to 28,000 lbf from the Pratt & Whitney's original TF-30 engine which produced 20,000 lbf of uninstalled thrust. The increase in thrust output of the GE engine increased the Tomcat's overall thrust-to-weight ratio while also increasing safety for carrier launches. The engine increased the 60% increased strike range for the F-14D variant along with a 61% increase in rate of climb. The GE's F110 engine also powered some F-15 Eagles (the F-15Ks exported to South Korea) and the F-16 (F100-GE-100 and F100-GE-129). The F-14B also received the ALR-67 Radar Homing & Warning System (RHAW). Around 38 new F-14Bs were produced while 48 F-14As were upgraded into B variants.

F-14D Super Tomcat: All F-14D variants, which entered service from 1991 onwards, were powered by the F110-GE-400 engines while also incorporating digital avionics, including, glass cockpit and newer AN/APG-71 radar which replaced the original AWG-9s. Other newer features on the D variant, included:-

- Airborne Self Protection Jammer (ASPJ)
- Joint Tactical Information Distribution System (JTIDS)
- SJU-17(V) Naval Aircrew Common Ejection Seats (NACES)
- Infrared Search and Track (IRST)
- Addition of LANTIRN Pods for Ground Strike Missions

Eventually, a total of 37 new F-14Ds were produced while 18 F-14As were upgraded to the D variant.

Primary Operators: U.S. Navy & Iranian Air Force

The U.S Navy: The Navy operated the F-14 from 1974 to 2006 with the leading user being the Navy Fighter Weapons School also known as Top Gun.

Iranian Air Force was the sole export customer & foreign operator of the Tomcat with deliveries beginning in the mid-1970s with Iran ordering 80 F-14 Tomcats with some of them delivered prior to the coup of 1979 in Iran. The Iranian F-14 Tomcats witnessed extensive usage with significant success against the Russian MiG-21s, MiG-23s and Su-20/22s during the Iran-Iraq war of the 1980s. Iran currently is the sole operator of the F-14 Tomcats in the world with an active, in-service fleet.

Specifications:

- **Crew:** 2
- **Length:** 62 ft 9 in (19.13 m)
- **Wingspan:** 64 ft 1.5 in (19.545 m)
- **Swept wingspan:** 38 ft 2.5 in (11.646 m) swept
- **Height:** 16 ft (4.9 m)
- **Wing Area:** 565 sq ft (52.5 m^2) wings only
- **Empty weight:** 43,735 lb (19,838 kg)
- **Gross weight:** 61,000 lb (27,669 kg)
- **Max takeoff weight:** 74,350 lb (33,725 kg)
- **Fuel capacity:** 16,200 lb (7,348 kg) internal fuel

Powerplant: 2 × General Electric F110-GE-400 afterburning turbofans producing 16,610 lbf (73.9 kN) thrust each dry and 28,200 lbf (125 kN) with afterburner

Performance

Top Speed: Mach 2.34

Ferry Range: 1600 NM

Combat Range: 500 NM

Service Ceiling: 53,000 ft

G Limits: +7.5

Rate of Climb: 45,000 ft./min

Wing Loading: 96 lb/sq. ft. (107.9 lb/sq. ft. calculated based on Gross Weight & Wing Area)

Thrust/Weight: 0.92 at Gross Weight (1.07 with loaded weight & 50% internal fuel

Data Sources: U.S. Navy Fact File on F-14 Tomcat, Spick, M.A.T.S., Flight International

Armament

- **Guns: 1× 20 mm (0.787 in) M61A1 Vulcan 6-barreled rotary cannon, with 675 rounds**
    - **Hardpoints:** 10 total with an overall payload capacity of 14,500 lb (6,600 kg)
    - **Rockets:** 7x LAU-10 rocket pods (for a total of 28 rockets)
    - **Missiles:** AIM-54 Phoenix, AIM-7 Sparrow, AIM-9 Sidewinder air-to-air missiles:
    - **Bombs:**
        - JDAM precision-guided munition (PGMs)
        - Paveway series of laser-guided bombs
        - Mk 80 series of unguided iron bombs
        - Mk 20 Rockeye II cluster munition
    - **Other:**

Tactical Airborne Reconnaissance Pod System (TARPS)

- LANTIRN Targeting System (LTS) pod (AN/AAQ-14)
- 2× 267 US gal drop tanks for extended range/loitering time

Avionics

- Hughes AN/APG-71 radar
- AN/ASN-130 Inertial navigation system, Infrared search and track, AAX-1 TCS

Program Status

The production of the F-14 Tomcat was shutdown in the 1990s with the in-service fleet decommissioned by the U.S. Navy in September 2006 with the same replaced by the F/A-18E/F Super Hornets (from 1999 onwards) developed & produced by McDonnell Douglas Corporation initially and later by Boeing following the acquisition of McDonnell Douglas by Boeing in 1998. Iranian Air

Force is the only active operator of F-14 Tomcat fleet which is over 50+ years old.

Image: An F-14 Tomcat flying in tandem with its successor the F/A-18 E/F Super Hornet

Image Credits: U.S. Navy, Lt. Cmdr. John Braun, Image ID: ID 050131-N-0000B-001, Public Domain

2. Boeing F-15E/EX Strike Eagle/ Eagle II

Description:

The F-15E Strike Eagle is the primary, multi-role, air superiority and strike fighter aircraft of the U.S. Air Force. The F-15E is based on the original McDonnell Douglas built F-15C/D Eagle built for air superiority role following the dogfighting debacles suffered in the Vietnam War. F-15EX Eagle II is the latest, 21st century incarnation of the F-15 platform.

Primary Role

The F-15E Strike Eagle is a twin engine, multirole strike aircraft of the U.S. Air Force and the program's primary role is to carry out long range, high-speed Air-to-Air interdictions without needing escort aircrafts or electronic attack aircrafts and strike ground targets. The F-15 program was originally conceived primarily as an Air Superiority aircraft with the program having proven itself as almost invincible in aerial dogfights with the F-15s having scored over a 100 aerial victories with not even a single one lost to the enemy.

Origin, Timeline & Scope

The F-15 Eagle program was originally conceived by the USAF as a replacement for its ageing, in-service fleet of F-4 Phantoms. The F-15 program was conceived in the late 1960s, under the USAF's plan to develop a dedicated air superiority fighter following heavy losses in dogfights through the Vietnam War. The McDonnell Douglas' design was down-selected in 1969 with the first prototype having its maiden flight in 1972 and the first F-15s entering service in 1976. The F-15E variant was developed following the proven success of the F-15 airframe in aerial warfare. The first F-15E had its maiden flight in 1986 and EIS two years later in 1988.

Quantities Produced & In-Service Fleet:

Overall, around 1,200 F-15s have been built by McDonnell Douglas since early 1970s and later by Boeing since 1997. Over 500+ F-15E aircrafts have been built since 1985.

Variants & Versions Built

F-15A: Single Seat, all-weather air-superiority fighter variant built from 1972-1979

F-15B: Two-seat Training variant

F-15C: Improved, Single Seat all-weather air-superiority fighter

F-15D: Two-seat training version

F-15J: Single Seat all-weather air-superiority fighter produced for Japan

F-15DJ: Two-Seat Trainer Variant produced for Japan

F15E Strike Eagle: Twin-Seat, all-weather, multi-role strike aircraft featuring CFTs. The F-15E was further developed into

- F-15I for Israel
- F-15SG for Singapore
- F-15K for South Korea
- F-15S, F-15SA for Saudi Arabia
- F-15EX for the USAF

Specifications: For F-15E Strike Eagle Variant

Crew: 2 (Pilot & WSO)

Length: 63 ft 9.6 in (19.446 m)

Wingspan: 42 ft 9.6 in (13.045 m)

Height: 18 ft 6 in (5.64 m)

Wing area: 608 sq ft (56.5 m2)

Wing Loading: 73.19 lb. per sq. ft.

Empty weight: 31,700 lb (14,379 kg)

Gross Weight: 44,500 lb (20,185 kg)

Max takeoff weight: 81,000 lb (36,741 kg)

Powerplant:

2 × **PW F100-PW-220** afterburning turbofans, producing 14,590 lbf each in dry mode and 23,770 lbf in wet mode (with afterburner)

Second-Engine Option

2XF100-PW-229: Thrust: 17,800 Dry and 29,160 Wet

Internal fuel capacity: 13,455 lb (6,103 kg)

Key Operators: The U.S. Air Force (USAF) has been the leading operator of the F-15E Strike Eagle followed by:-

- Israel

- Saudi Arabia
- Qatar
- South Korea
- Singapore

Performance:

Maximum speed: Mach 2.5 (1,650 mph, 2,655 km/h) at High Altitude

Maximum speed at Sea Level: Mach 1.2

Combat range: 687 NM (791 mi, 1,272 km) for interdiction

Ferry range: 2,100 NM (2,400 mi, 3,900 km)

Service ceiling: 60,000 ft (18,000 m)

Rate of climb: 50,000 ft/min (250 m/s)

Thrust/weight: 1.005

G Limits: +9G

Data Sources: USAF Fact File, International Directory of Military Aircraft, Encyclopedia of Modern Military Aircraft, Jane's All the World's Aircraft, USAF F-15A/B/C/D Manual

Image Credits: USAF

Armament & Avionics:

The F-15 airframe has 9 Hard points with a total payload capacity of 23,000 lb./10,500 kg. of External Fuel & Ordnance. The armaments which can be carried by the F-15 include:-

Guns

1X20mm M61A1 Vulcan 6-barreled Gatling Cannon

Missiles:

4× AIM-9 Sidewinder - short range air-air missile

8× AIM-120 AMRAAM - medium range air-air missile

4× AIM-7 Sparrow - medium range air-air missile

Air-to-Surface Missiles:

6× AGM-65 Maverick

2× AGM-84 Harpoon

2× AGM-84H/K SLAM-ER

AGM-130

AGM-154 JSOW

AGM-158 JASSM

Others:

- Up to 3 External Drop Fuel Tanks capable of carrying 600 US Gallons of Fuel for Extended Operating Range/ Loitering Time

- 1 Super Cruise Drop Tank Capable of carrying 480 US Gallon of Fuel

Avionics:

Radar

- Raytheon AN/APG-70 or AN/APG-82

- AN/ASQ-236 Radar Pod

Targeting Pods

LANTIRN or Lockheed Martin Sniper XR or LITENING targeting pods

Countermeasures:

- Northrop Grumman Electronic Systems AN/ALQ-131 electronic countermeasures pod
- Hazeltine AN/APX-76 or Raytheon AN/APX-119 Identify Friend/ Foe (IFF) interrogator
- Magnavox AN/ALQ-128 Electronic Warfare Warning Set (EWWS) – part of Tactical Electronic Warfare Systems (TEWS)
- Loral AN/ALR-56 Radar warning receivers (RWR) – part of TEWS
- Northrop Grumman Electronic Systems ALQ-135 Internal Countermeasures System (ICS) – part of TEWS

- Marconi AN/ALE-45 Chaff/Flares dispenser system – part of TEWS

Program Status:

Currently in production as well as in-service and amongst the world's leading twin-engine, multi-role Air Superiority fighter jet aircraft programs

OEM – The Boeing Company

Boeing produces the F-15E aircraft program at its St. Louis, Missouri facility

Key Mission Profiles

- Multi-Role Aircraft Program – Air Superiority Fighter (F-15 A/B/C/D Variants)

- Air-to-Air Interdiction

- Ground Attack (F-15E Strike Eagle Variant)

The F-15EX

F-15EX is a single seat variant aimed at replacing the USAF's existing, in-service F-15C/Ds for airbase & homeland defense role given the delay in deliveries of F-35s. The F-15EX includes AMBER weapons rack enabling it to carry 22 air-to-air missiles, Infrared Search & Track Radar, advanced avionics in addition to electronic warfare equipment, AESA radar and extended service life of up to 20,000 hours.

Key Recent Contract Awards

- Boeing received a multi-year contract award from the U.S. Air Force in July 2020 for procurement of 8 F-15EX fighter jets through 2023.

- The multi-year contract is worth $1.1 billion

- Boeing delivered the first F-15EX to the USAF in March 2021 with the program making its maiden flight in February 2021

- USAF plans to procure a total of 144 F-15EXs going forward at a URF of $87.7 million which values the F-15EX procurement program at $23 billion.

Proposed Upgrades & Strategy Focus

Boeing has incorporated the following upgrades into the F-15EX for the USAF:

- AMBER Weapons Rack

- AESA Radar

- Infrared Search & Track Radar

- Extended Service Life of 20,000 Hours

- Open Mission Systems Architecture enabling Future Upgrades

F-15 Silent Eagle (SE)

Boeing had also proposed a F-15SE Silent Eagle variant for Air-to-Air missions which offered some 5th generation features, including:-

Conformal Weapons Bay (CWB) for Internal Storage of Weapons

Usage of Radar-Absorbent Materials and Outward Canting of Twin-Tales for Reduced Radar Cross-Section

The First F-15SE demonstrator took-off in 2010 and Boeing had offered it to Israel, Saudi Arabia, Japan and South Korea (all traditional F-15E operators).

However, all of them, except Saudi Arabia, zeroed in on & selected the 5th generation F-35 program while Saudi Arabia went for the F-15SA. The F-15EX will be powered by the F110-GE-129 engines. Focus is on getting some of the earlier users of the F-15E across international markets to the latest

F-15EX program while dissuading them from going for the F-35 which has eaten up a lot of market share from Boeing for legacy Hornet fleet replacement globally.

F-15E/EX

Strengths:

1. F-15E has been the USAF's top of the line air superiority, heavy fighter jet with tremendous maneuverability which has made the F-15 virtually almost invincible in the Aerial combat and dogfights over the decades with its unbeatable track record.

1. The F-15E has one of the highest payload capacity among all the 6 fighter jet programs (at 23,000 lb) enabling it to carry the largest and most broad range of weapons, armaments & munitions. The only aircraft which comes closest to matching the payload capacity of F-15E is Dassault Rafale with a payload capacity of 20,900 lb. The Rafale's MTOW, however, is almost 50% less than the F-15E's gigantic 81,000 lb MTOW.
2. The F-15E is also the fastest of the six fighter jet aircraft programs being compared with a top speed of Mach 2.5 at high altitude. The Eurofighter Typhoon & JAS 39 Gripen reach the Mach 2 mark while the F-16 has Mach 2.05 as its high altitude top speed.
3. The F-15E forms the backbone of USAF's force structure and Air Defense capabilities and is likely to stay that way for a significant time to come given its resurgence as the F-15EX, with USAF placing fresh orders, as a replacement for the older F-15C/D variants.
4. Very high reliability of the F-15E program with a loss of aircraft registered only at almost 50,000 flying hours.
5. Very cost effective Total Cost of Ownership (TCO) proposition for the F-15EX which is equivalent of procuring & sustaining almost 3 F-35s on a comparative basis.
6. High average life span of almost 20,000 flying hours for the latest iteration of the F-15, the F-15EX.
7. Effective engine redundancy with the availability of twin engines

onboard.

8. Huge installed, in-service fleet of F-15 program with 500+ F15E variants and around 1,200 total F-15 aircrafts in service across Air Forces globally.

Weaknesses:

- Comparative procurement cost on a URF unit basis seems to be the downside for the latest variant of the F-15 program, the F-15EX. For instance, the URF for the 4.5 generation F-15EX (devoid of any stealth or sensor fusion capabilities) being procured by the USAF comes out to be $87.7 million which clearly is even higher than the F-35's latest URF cost of $77.9 million. The upfront costs seem to be the factor which could tilt the scales in favor of the modern, 5th generation F-35.

- The F-15EX is ineffective in operating in contested airspaces and can not penetrate modern air defenses and its application, thus, is limited to the homeland defense and air interdiction roles with its ground strike role likely to be limited comparatively.

- The F-15 platform is devoid of canards or thrust vectoring capability (TVC) and rather leverages its low wing loading for maneuverability which could impact its dogfighting ability in within-visual-range engagements against much more maneuverable aircrafts, like the Eurofighter Typhoon, Rafale or the Russian Su-30Flanker/Su-35 Super Flanker family, which the Su-30/35 being two of the most maneuverable fighter jets (besides the F-22 Raptor) featuring canards as well as TVC capability.

- The F-15 program lacks many aerodynamic enhancements which are routine in many 3rd & 4th generation aircraft programs, like, relaxed static stability, lifting body configuration, wing leading edge slats and wing root extensions. Also, on the F-15 airframe, there is a trade-off between payload and maneuvering ability given that it derives its lift & maneuverability primarily from its low wing loading.

- The F-15EX is likely to be limited to the U.S. market, primarily, as it may not be very competitive from a quantitatively perspective while competing against other 4.5 or 5th generation aircraft programs from competitors on a TCO basis given that the F-15 is a heavy fighter jet featuring previous generation engines.

- F-15EX's high operating cost per flight hour for a 4.5 generation program standing at almost $29,000 per hour. The F-35 comparatively costs almost $44,000 per hour to operate.

Maverick's Choice - F-15E Strike Eagle vs. F-14D Super Tomcat – Face-Off

The F-15 Eagle and the F-14 Tomcat airframes belonged to the same development era of late 1960s and early to mid 1970s dominated by the Vietnam War's aftershocks and the setback faced by U.S. fighter Jets, especially, F-4 Phantoms against the Russian jets in dogfights operated by the North Vietnamese. The reason for the series of U.S. setbacks was the adoption of operational doctrine of stand-off weapon systems in the age of air-to-air missiles which theoretically obviated the need for close range dogfighting capabilities for the fighter jets as well as pilots. This realization during the war led to a fundamental shift in the conceptualization, design & development of fighter jets in the 1970s based on actual, aerial combat experience of pilots with focus on a balance of high maneuverability, armament payload, speed and rate of climb. The fighter mafia of the 1970s, which played a pivotal role in the design & development of the F-16 platform under Colonel John Boyd and his revolutionary Energy Maneuverability (E-M) theory, for the USAF's Lightweight Fighter (LWF) Program, is a prominent example.

The F-15 Eagle and the F-14 Tomcat have strikingly similar technical specifications owing to the purpose and the era in which they were built at the peak of Cold War, wherein, absolute technological overmatch was a clear imperative and sure advantage in the skies over the adversary.

Let's compare the F-15E Strike Eagle and the F-14D Super Tomcat in terms of tech specifications:-

Parameters	F14D vs. F15E	
	F14D Tomcat	F15E Strike Eagle
Empty & Gross Weight	43,735 lb/61,000 lb	31,700 lb/58,000 lb
MTOW	74,350 lb	81,000 lb
Internal Fuel Capacity	16,200 lb	13,455 lb
Total Engine Thrust (Dry/Wet)	16,610 lbf/28,200 lbf	14,590 lbf/23,770 lbf (For F100-PW-220) 17,800 lbf/29,160 lbf (for F100-PW-229)
Top Speed	Mach 2.34	Mach 2.5+
Combat/Operating Range	500 NM/1600 NM	687 NM/2,100 NM
Service Ceiling	53,000 ft.	60,000 ft.
Rate of Climb	45,000 ft/min	50,000 ft./min
Wing Loading	96/107.9 lb/sq. ft.	73.19 lb/sq. ft.
Thrust to Weight Ratio	0.92	1.005 (F100-PW-229)
Hard Points	10	11
Payload Capacity	14,500 lb	23,000 lb

The first & most striking point to be noted is that the F-14D is a slightly heavier aircraft than the F-15E with the Tomcat's gross weight being 61,000 lb against the Eagle's 58,000 lb. Taking-off and landing such a heavy fighter on the carrier deck would have been no easy feat for the pilots and the carrier crew with the problems further compounded for the initial F14As which were powered by the Pratt & Whitney's TF30 turbofans with their limited thrust and stalling issues which necessitated a full-afterburner take-off leading to a very high fuel consumption right at the outset constraining the effective operating & combat range without refueling. The issues were resolved following the incorporation of F110-GE-400 turbofans onto the F14B Tomcat variant with the TF30 effectively replaced.

The F15E's airframe has a 10% higher MTOW than the F14D Tomcat which enables the Eagle to carry a 33% more armament payload than the Tomcat with the Eagle's armament payload being a massive 23,000 lb and that's why the Israeli Air Force has given F15E the moniker, 'Flying Bomb Truck'.

Another key parameter, pertaining to maneuverability, however, also favors the F15E with the Eagle's wing loading being much lower at 73 lb/sq. ft. as compared against the F14D Tomcat's much higher wing loading number of around 96/107 lb/ sq. ft. given the Tomcat's variable sweep wing design (with a sweep range of 20 degrees to 68 degrees with the high sweep angle boosting high speed performance significantly) which, in turn, provides a much lower wing loading than the numbers suggest with the fuselage acting as a lifting surface and providing almost 40% to 60% of overall lift. Additionally, the F-15 Eagle, too, has that almost blended wing-body design with the fixed wing mounted on to a flat, wide fuselage which provides additional lift by acting also as a lifting surface just like the F-14 Tomcat. When the wing loading is combined with the thrust-to-weight ratio of around 0.92 for Tomcat and 1.005 for the Eagle along with its cropped delta wing, the F15E clearly comes on top with its much lower wing loading, higher thrust-to-weight ratio giving it better maneuverability along with a 10% faster vertical climb rate. Overall, better combat and operating ranges, a relatively higher top speed and higher payload capacity clearly skew the head-to-head comparison between the two towards the F15E Strike Eagle.

The F-15 Eagle, thus, is virtually indomitable in close aerial combat or dogfights (especially the Single Seat A/C Air Superiority variants) given that it can pull off up to 9Gs, which a skilled pilot can really exploit to the hilt for a definitive advantage, while the Tomcat can do a max 6.5 or 7Gs. The Tomcat, on the contrary, is great for tracking & engaging multiple targets simultaneously with its powerful AWG-9 radar and shooting them down from beyond-visual-ranges with its AIM-54 Phoenix missiles which the Eagle clearly lacks as it has been designed & custom built for dogfights.

The prospect of flying the F-15E, thus, would surely have put a smile on Maverick's face! However, given the fact that the F-15E Strike Eagle can't operate from the carrier deck as it has not designed for it and it would rather would have had to be operated from land would have been a dampener.

In fact, McDonnell Douglas did pitch the Navy for a naval variant of F-15N, dubbed as the Sea Eagle, in 1971 which was, in turn, turned down by the Navy over weight issues, as unraveled by author Robert S. Hopkins III in his book, "McDonnell Douglas F-15 Eagle: Supreme Heavy-Weight Fighter"*[1]

The incorporation of carrier-specific modifications, including, a reinforced landing gear, a stronger arrester tail hook and folding wings etc. would have increased the gross weight of the proposed F-15N by around 3,000 lb. than the land based variants of the F-15 which would have pegged the gross weight of the F-15N exactly equal to the F-14 Tomcat's at 61,000 lb and would have impacted the thrust-to-weight ratio and maneuverability for Aerial Combat Maneuvers (ACM).

Further, arming the F-15N with the AIM-54 Phoenix missiles as well as the AWG-9 radar (as the F-15 at the time was only armed with Sparrow & Sidewinders), to create the F-15N-PHX, in order to match the Tomcat's capability to track & engage multiple targets simultaneously from Beyond-Visual-Range (BVR), just like Tomcat, would have increased the F-15 Eagle's gross weight by almost 10,000 lb, as compared to the conventional F-15A and, thus, would have rendered the entire exercise pointless as it would not have offered any advantage to the F-15N-PHX over Tomcat, except, in reduced maintenance costs owing to F-14 Tomcat's swing wing design.

The F14 Tomcat and the F-15 Eagle, thus, have been two of the most elegant, best & unmatched fighter jet aircrafts in military aviation.

3. Boeing F/A-18E/F Super Hornet (Nickname: Rhino)

Image Credits: Pixabay.com

Description:

The F/A-18E/F is the primary, multi-role combat aircraft of the U.S. Navy capable of operating from aircraft carriers. In its second generation since 1999, the F/A-18E/F is based on the original McDonnell Douglas built F/A-18 Hornet and has replaced the U.S. Navy's F-14 Tomcat fleet.

Primary Role

The F/A-18E/F is a twin engine, multirole primary strike aircraft of the U.S. Navy and the program's primary role is to carry out carrier based interdiction & strike missions. Other missions profiles, include, Fighter Escort, Fleet Air Defense, Suppression of Enemy Air Defenses, Air Interdiction, Close Air Support, Aerial Reconnaissance and Refueling based on External Air-To-Air Refueling System

Origin, Timeline & Scope

The F/A-18E/F program was originally conceived as an evolution of the original F/A-18 Hornet with a comprehensive re-design. In early 1990s, post the retirement of the F-14 Tomcat fleet, the Navy focused on reducing the aircraft types it operates and focused on increasing the procurement of Super Hornets to replace multiple aircraft types in its fleet, including, the A-6 Intruder, Lockheed S-3 Viking & the KA-6D aircrafts. The Super Hornet is almost 20% larger than the original Hornet, features 33% larger internal fuel carrying capacity, a 41% enhanced mission range and a 50% increase in endurance. The Super Hornet had its maiden flight in 1995 and entered service with the U.S. Navy in 1999.

Quantities Produced & In-Service Fleet: :

- Over 600+ F/A-18E/F aircrafts have been built over the 1999-April 2020 period, as per U.S. Navy.

Variants & Versions Built

F/A-18E Super Hornet: single seat variant.

F/A-18F Super Hornet: two-seat variant.

EA-18G Growler: Electronic warfare variant of the F/A-18F as a replacement for the U.S. Navy's Northrop Grumman EA-6B Prowler.

Advanced Super Hornet: Variant of the F/A-18E/F Super Hornet featuring Conformal Fuel Tanks (CFT), a further reduced radar cross section (RCS) along with the option of a stealthy enclosed weapons pod and built-in IRST21 sensor system.

Image Credits: Pixabay.com

Specifications:

Crew: 1 for F/A-18E and 2 for F/A-18F

Length: 60 ft 1.25 in (18.31 m)

Wingspan: 44 ft 8.5 in (13.62 m)

Height: 16 ft 0 in (4.88 m)

Wing area: 500 sq ft (46.5 m2)

Empty weight: 32,081 lb (14,552 kg)

Gross Weight: 47,000 lb (21,320 kg)

Max takeoff weight: 66,000 lb (29,937 kg)

Powerplant: 2 × GE F414-GE-400 turbofans, producing 13,000 lbf each in dry mode and 22,000 lbf in wet mode

Internal fuel capacity: F/A-18E: 14,700 lb (6,667 kg), F/A-18F: 13,760 lb (6,241 kg)

External fuel capacity: Up to 4 × 480 gal tanks, totaling 13,040 lb (5,914 kg)

Key Operators: The U.S. Navy has been the leading operator of the F/A-18E/F Super Hornet

followed by the Royal Australian Air Force (RAAF) and the Kuwait Air Force which has 28 aircrafts on order.

Performance

Maximum speed: 1,030 kn (1,190 mph, 1,915 km/h) at 40,000 ft (12,190 m)

Maximum speed: Mach 1.8+

Combat Range: 1,275 nmi (1,458 mi, 2,346 km) clean plus two AIM-9s

Ferry range: 1,660 nmi (2,070 mi, 3,330 km)

Service ceiling: 50,000+ ft (15,000 m) at least

Rate of climb: 44,882 ft/min (228 m/s)

Wing loading: 94.0 lb/sq ft (459 kg/m2)

Thrust/weight: 0.93

Design load factor: 7.6 g

Armament & Avionics:

11 Hard points with a total payload capacity of 17,750 lb. The armaments which can be carried by the F/A-18E/F include:-

Guns

20mm M61A2 Vulcan Gun

Missiles:

4× AIM-9 Sidewinder - short range air-air missile

12× AIM-120 AMRAAM - medium range air-air missile

Sources: U.S. Navy Fact File, International Directory of Military Aircraft, Encyclopedia of Modern Military Aircraft

4× AIM-7 Sparrow - medium range air-air missile

6× AGM-65E/F Maverick - air-to-ground missile

4× AGM-84H/K SLAM-ER - air-to-ground cruise missile

6× AGM-88 HARM - anti-radiation missile

4× AGM-154 Joint Standoff Weapon - glide bomb

AGM-158 JASSM - air-to-surface missile

2× AGM-84 Harpoon - anti-ship missile

Bombs:

JDAM bombs (up to 10× GBU-32/35/38/54 or 4× GBU-31)

Paveway series of laser-guided bombs

Mk 80 series unguided bombs

CBU-78 Gator mine system

Mk 20 Rockeye II cluster bomb

Mk-62/63/65 Quick Strike Naval mine

Others:

- SUU-42A/A Flares/infrared decoy dispenser pod and chaff pod

- AN/ALE-50 towed decoy system pod

- AN/ASQ-228 ATFLIR Targeting pods up to 4× 480 US gal (1,800 l; 400 imp gal) drop tanks and 1× A/A42R-1 Aerial Refueling Store pod for aerial refueling.

- 12× ADM-141C TALD decoys

- AWW-13 Advanced data link pod

Sources: U.S. Navy Fact File, International Directory of Military Aircraft, Encyclopedia of Modern Military Aircraft

Avionics:

- Hughes APG-73 or Raytheon APG-79 Radar

- Northrop Grumman/ITT ALQ-165 self-protection jammer system or BAE Systems AN/ALE-214 integrated defensive electronic countermeasures system

- Raytheon AN/ALE-50 or BAE Systems AN/ALE-55 towed decoy

- Raytheon AN/ALR-67(V)3 radar warning receiver

- MIDS LVT or MIDS JTRS datalink transceiver

Program Status:

Currently in production as well as in-service and amongst the world's leading twin-engine, multi-role fighter jet aircraft programs

OEM – The Boeing Company

Boeing produces the F/A-18E/F aircraft at its St. Louis, Missouri facility

Key Mission Profiles

- Multi-Role Aircraft Program – Aerial Defense & Close Air Support

Variants

- F/A-18E Super Hornet: Single Seat Variant

- F/A-18F Super Hornet: Two-Seat Variant

• EA-18G Growler Electronic Warfare Aircraft

• Advanced Super Hornet: Variant featuring Conformal Fuel Tanks (CFTs) and a further reduced Radar Cross Section (RCS) apart from stealthy, enclosed weapons pod and an in-built IRST21 sensor system

Key Recent Contract Awards

• Boeing received a multi-year contract award from the U.S. Navy in 2019 for 78 F/A-18E/F fighter jets as replacement for its ageing, in-service fleet of F/A-18C-D aircrafts and as an interim measure to compensate for the slow deliveries of the F-35C Lightning II program

• The multi-year contract is worth $4 billion and involves deliveries of the aircrafts by 2021. Boeing estimates that the award of multi-year contract would save the Navy almost $395 million.

• Boeing estimates that the Navy plans to buy around 110 F/A-18E/Fs to replace its legacy fleet of F/A-18A-D aircrafts, including, the latest multi-year contract award for 78 aircrafts.

Proposed Upgrades & Program Strategy Focus

1. Boeing proposed a new Advanced Super Hornet in early 2010s which were to feature:-

• 50% reduced Frontal Radar Cross Section (RCS) for a relatively lower observability

• Conformal Fuel Tanks (CFT) to carry 3,500 lb. of additional fuel which would increase the operating range of the Super Hornet by 260 NM

• Enclosed Weapons Pod (EWP)

1. Boeing and Northrop Grumman self-funded the development of a prototype which underwent flight testing in March 2013.
2. Boeing also proposed a Super Hornet Hybrid Concept in 2014 featuring incorporation of electronic signal detection capabilities derived from the EA-18G Growler which would have enabled target engagement using the receiver. The new Hybrid concept could also incorporate Long Range Infrared Search and Track Sensor and Air-to-Air Tracking Modes. The strategy focus of the program for Boeing clearly is on pursuing legacy F-18 Hornet replacement contests across key global markets.

F/A-18E/F Super Hornet:

Strengths:

- The F/A-18 platform's strategic positioning as the primary, multi-role fighter aircraft of the U.S. Navy and a number of allied Air Forces, primarily, Australia, Canada, Finland, Spain, Switzerland, Kuwait and Malaysia.

- Significant global installed & in-service fleet of F/A-18s numbered at around 2,000 aircrafts with 500+ F/A-18E/F Super Hornets and around 1,480 F/A-18 Hornets distributed across the U.S. Navy and allied forces.

- High reliability and versatility of the F/A-18 aircraft platform with its highly aerodynamic characteristics, avionics and cockpit displays. Additionally, the F/A-18 has proved itself in active duty & combat service over the past 2+ decades

- The ready availability of the F/A-18E/F as a 4th generation, proven & cost effective aircraft platform available at almost half the cost of a 5th generation aircraft program, like the F-35 Lightning II program which is struggling to keep URF cost at under $100 million.

• The already significant installed, global base of older F/A-18C/D variants across multiple, allied nations, at around 1,480 aircrafts globally, provides Boeing with a ready user base to pitch replacements in form of the latest F/A-18E/F leveraging a ready, in-country sustainment support infrastructure in-place providing advantages with lower operating costs & an overall favourable TCO equation.

• The commonality of the F/A-18 airframe with the EA-18G Growler Electronic Warfare aircraft Variant provides a significant incentive to nations already operating the older Hornets to continue with the latest F/A-18 Super Hornets as replacements for the older Hornets.

Weaknesses:

• The airframe design of the F/A-18 is over 50 years old given that it was based on the YF-17 designed by Northrop Grumman in the 1970s for the U.S. Navy and the US Marine Corps.

• The F/A-18 program has faced criticism for its limited payload capability and operating range as compared to its predecessors, especially, the venerable F-14 Tomcat built by Northrop Grumman Corporation. The F/A-18E/F nears the payload and operating range figures of the F-14 Tomcat but no match for the old master.

• The US Navy & the USMC have observed & have been battling structural corrosion damage & issues, more rapid than had been anticipated, over its in-service fleet of legacy F-18s given their extended service horizon from 6,000 hours earlier to 10,000 hours and the prolonged exposure to the oceans' salt water environment. The same has led to accelerated retirement of legacy F-18s which is being replaced by latest procurement orders for about 100 F/A-18E/ Fs by the Navy.

• A number of allied nations operating the legacy F/A-18C/D legacy Hornet fleets are now looking for replacement with some of them moving towards & switching to the 5th generation F-35s instead of choosing & moving to the 4th generation F/A-18E/F variants. For instance, Poland & Switzerland have already chosen the F-35As as the replacement for its legacy F-18 & F-5 fleets and placed order for 36 F-35s. Finland & Canada, too, have gone for the F-35as a replacement for their fleets of legacy F-18 & CF-18 Hornets joined by Germany which also has chosen the F-35 with the global market for fighter jets really heating up following the Russia-Ukraine fighter with around 400+ fighter jets likely to be delivered globally over the next decade and it is going to be an all out war between the bevy of American & European fighter jet manufacturers for a share of the global pie.

Parameters	F14D vs. F/A-18 E/F	
	F14D Tomcat	F/A-18E/F Super Hornet
Empty & Gross Weight	43,735 lb/61,000 lb	32,081 lb/47,000 lb
MTOW	74,350 lb	66,000 lb
Internal Fuel Capacity	16,200 lb	14,700 lb (E)/13.760(F)
Total Engine Thrust (Dry/Wet)	16,610 lbf/28,200 lbf	13,000 lbf/22,000 lbf
Top Speed	Mach 2.34	Mach 1.6
Combat Range/Range	500 NM/1600 NM	390 NM/1275 NM
Service Ceiling	53,000 ft.	50,000 ft.
Rate of Climb	45,000 ft/min	44,882 ft./min
Wingspan	64 ft. 1.5 in	44 ft. 8.5 in
Wing Loading	96/107.9 lb/sq. ft.	94 lb/sq. ft.
Thrust to Weight Ratio	0.92	0.93
Hard Points	10	11
Payload Capacity	14,500 lb	17,750 lb

The eventual successor to the F-14 Tomcat on the Navy's carrier decks, the F/A-18E/F Super Hornet, is a much lighter bird than the Tomcat with a 25% lower empty and 23% lower gross weight along with a 11% lower MTOW of 66,000 lb than the Tomcat's 74,350 lb. The Tomcat also is identical to the Super Hornet in length and height. However, with a much larger wing with almost a 50% larger wingspan at 64 ft. 1.5 in as compared to Super Hornet's 44 ft. 8.5 in. The Super Hornet also has been produced in a single seat variant needing only the pilot unlike the Tomcat which had been custom-built as a twin-seat aircraft needing a dedicated Radar Intercept Officer (RIO) at the back which made virtually Goose indispensable for Maverick in the cockpit.

The Super Hornet's internal fuel carrying capacity is also around 9% lower than its predecessor along with a 20% lower combat range of 390 NM as against

Tomcat's 500 NM. However, the biggest let down for Maverick here would be the Super Hornet's much lower top speed of Mach 1.6 as against Tomcat's top speed of Mach 2.34 which effectively jettisons the Super Hornet off the carrier decks for Maverick on his latest revenant given his need for speed. The payload capacity for the Super Hornet, surprisingly, is 15% higher than the Tomcat given its lower empty & gross weight as well as lower MTOW and despite reduced engine thrust.

The remaining parameters, including, service ceiling, vertical climb rate, wing loading and the thrust-to-weight ratio are almost similar between the two platforms with McDonnell Douglas cleverly keeping the Super Hornet's tech specs similar to the outgoing Tomcat as much as possible while designing it and reducing the speed and engine thrust for efficiency given that the Super Hornet is much more balanced has been designed for a much broader role & set of mission profiles, while preserving its fighter & self-defense capabilities, which are way superior to the Tomcat whose primary role was air-to-air interdiction of Russian bombers, fighters & subs at high supersonic speeds as the spearhead of carrier strike group defense at the peak of Cold War.

The Super Hornet, thus, effectively has traded power for agility and thus misses the Tomcat's raw muscle power and that would have been a big disappointment and a big no for Maverick!

3. Lockheed Martin F-16V Fighting Falcon Block 70

Image Credits: Pixabay.com

Description:

The F-16V Fighting Falcon is a single engine, multi-role fighter aircraft built originally by General Dynamics in the mid-1970s for the USAF. The latest iteration of the F-16 has been the F-16V Block 70.

Primary Role

The F-16 was originally conceived as a an air superiority day fighter. However, it evolved into an all weather multi-role aircraft. The F-16 was the world's first fully fly-by-wire fighter aircraft and featured side stick controls.

Origin, Timeline & Scope

The program was conceived originally in the late 1960s under the Light Weight Fighter (LWF) Program to develop a lightweight, cost effective fighter jet program custom-built for air superiority and Air-to-Air combat missions based on Col. John Boyd and his Fighter Mafia's Energy-Maneuverability theory. The program was also eyeing the market for replacement of NATO Allies' F-104G Starfighter aircraft fleet while the USAF itself needed to replace its fleet of F-105 and F-4 Phantom aircrafts. The General Dynamics built YF-16, powered by the PW100 Turbofan, was adjudged as the winner of the LWF program in 1975 while the competing design from Northrop & McDonnell Douglas team, the YF-17, was taken-up by the Navy and it ultimately transformed into the F/A-18 Hornet. The USAF planned to procure almost 650-1400 F-16 aircrafts. The first aircraft rolled out in 1976 and entered service in early 1979.

Quantities Produced & In-Service Fleet:

- Almost 4,600+ F-16 aircrafts so far with the final assembly line still operational after almost 47 years of commencement of production in 1973. In 1990s the F-16 program was sold by General Dynamics to Lockheed Martin Corporation.

Variants & Versions Built

F-16A/B: Initial Production Versions with the A being Single Seat variant while the B being 2-seat variant.

F-16C/D: Featured improved avionics and radar providing All Weather Capability and Beyond Visual Range AIM-7 & AIM-120 missiles. Production of C/D variants started in 1984.

F-16E/F: E was single seat variant while F was twin seat variant. The E/F featured an advanced AESA radar, advanced Avionics, Conformal Fuel Tanks (CFTs) and higher thrust GE engine, namely, F110-GE-132 engine.

F-16V: Latest version currently in production with the V standing for the F-16's nickname, Viper. The F-16V features an APG-83 AESA radar, new mission computer, electronic warfare suite, Lockheed Martin built Automated Ground Collision Avoidance System (AGCAS) and other avionics improvements to the cockpit. Lockheed Martin also targeting the F-16 upgrade market with this upgrade kit which can retrofitted on older F-16s.

Specifications:

<u>General characteristics</u>

Crew: 1

Length: 49 ft 4 in (15.06 m)

Wingspan: 32 ft 8 in (9.96 m)

Height: 16 ft (4.9 m)

Wing area: 300 sq ft (28 sq. m.)

Empty weight: 18,900 lb

Gross weight: 26,500 lb.

Max takeoff weight: 42,300 lb.

Fuel capacity: 7,000 pounds (3,200 kg.) internal fuel

Powerplant:

1 × Pratt & Whitney F100-PW-229 Afterburning Turbofan producing 17,800 lbf thrust in dry mode and 29,560 lbf in wet mode.

1X GE F110-GE-129 Afterburning Turbofan producing 17,155 lbf thrust in dry mode and 29,500 lbf thrust in wet mode.

Performance

Maximum Speed: Mach 2.05 (1,318 mph) at 40,000 feet

Mach 1.2 at Sea Level

Combat Range: 295 NM (339 miles or 546 km) with 4 X 1,000lb Bombs

Ferry range: 2,277 NM (2,620 miles or 4,217 km) with drop tanks

Service ceiling: 60,000 ft

G Limits: +9G

Rate of Climb: 72,000 ft/min

Thrust/Weight Ratio: 1.095

Roll Rate: 324 degrees/second

Key Operators:

USAF is the Primary Operator. 25 other Nations have been operating the F-16 aircraft. The Global In-Service fleet of the F-16 program stands at over 4000+ aircrafts.

Armament & Avionics:

11 Hard points with a total payload capacity of 17,000 lb. of External Fuel & Ordnance. The armaments which can be carried by the F-16 include:-

Guns

1X20mm M61A1 Vulcan 6-barreled Gatling Cannon

Missiles:

Air-to-Air Missiles

6 x AIM-9 Sidewinder - short range air-air missile

6 x AIM-120 AMRAAM - medium range air-air missile

6 x IRIS-T

6 x Python-4

6 x Python-5

AIM-7 Sparrow medium range AAMs

Air-to-Surface Missiles:

6× AGM-65 Maverick

2× AGM-88 HARM

4 x AGM-154 JSOW

AGM-158 JASSM

Anti-Ship Missiles

2 x AGM-84 Harpoon

4 x AGM-119 Penguin

Others: Up to 3 Drop Tanks carrying up to 600 US Gallon Fuel for Extended Range/Loitering

Avionics:

Radar

- Raytheon AN/APG-68 Radar
- MIL-STD-1553 Bus

Targeting Pods

LANTIRN or Lockheed Martin Sniper XR or LITENING targeting pods

Bombs:

8 × CBU-87 Combined Effects Munition

8 × CBU-89 Gator mine

8 × CBU-97 Sensor Fuzed Weapon

4 × Mark 84 general-purpose bombs

8 × Mark 83 GP bombs

12 × Mark 82 GP bombs

8 × GBU-39 Small Diameter Bomb (SDB)

4 × GBU-10 Paveway II

6 × GBU-12 Paveway II

4 × GBU-24 Paveway III

4 × GBU-27 Paveway III

4 × Joint Direct Attack Munition (JDAM) series

Wind Corrected Munitions Dispenser (WCMD)

B61 nuclear bomb

B83 nuclear bomb

Program Status:

Currently in production as well as in-service as amongst the world's longest running fighter jet aircraft program

OEM – Lockheed Martin Corpration

LMT has been producing the F-16 aircraft program at its latest production facility based in Greenville, South Carolina [1] for International Customers by moving the line from Fort Worth, Texas to make space for the F-35 program. The latest facility at Greenville opened in 2019 on Veterans Day.

Key Mission Profiles

- Multi-Role Aircraft Program – Air Superiority Fighter

- Air-to-Air Interdiction & Combat

- Ground Attack

---

1. https://www.af.mil/News/Article-Display/Article/2619834/air-force-opens-new-f-16-production-line-for-foreign-military-sales/

Key Recent Contract Awards & Other Developments

- Lockheed Martin has been producing the F-16V at its Greenville, SC based new facility since 2019 for International Exports customers. The company has witnessed a recent surge in orders for the F-16 given a rapid change in traditional world order and geopolitical conflict across parts of the world.

- Lockheed Martin received order for 19 F-16V fighter jets from Bahrain worth $2.8 billion apart from modernization & upgrade of Greece's 123 F-16s worth $2.4 billion[2]. Lockheed Martin had around 325 F-16 modernization contracts in hand worth $6.5 billion as of 2017 with more in the pipeline.

- The USAF's Tactical Aircraft Fleet Review points towards glaring capacity shortfalls which could be bridged most prudently by ordering in-production aircraft. This is what led to fresh orders for the F-15EX for Boeing from the USAF. On a similar line, USAF is also mulling plans to place fresh orders for the F-16 Block 70/72 after almost 2 decades, as reported by Aviation Week[3] in early January 2021.

- Lockheed Martin already has orders for F-16 in its order book from nations including, Slovakia, Bulgaria and Taiwan worth billions of dollars. The latest assembly line at Greenville will start rolling out F-16s from 2022 which will be delivered to these nations.

- Looking at Lockheed Martin's order book for the F-16 program and the sizeable market for upgrades of older F-16C/D variants, it becomes clear that the F-16 program is going to be one of the longest running fighter jet programs in the world with F-16s likely to be visible in the skies even beyond 2050. Lockheed Martin is looking

---

2. https://www.business-standard.com/article/companies/lockheed-martin-says-f-16-orders-flowing-in-117121600953_1.html

3. https://aviationweek.com/defense-space/aircraft-propulsion/us-air-force-talks-new-f-16-orders-latest-acquisition-shake

to pitch the latest F-16V to emerging markets while effectively harnessing the upgrade & modernization of legacy F-16 fleet for itself.

1. **F-16**
2. **Strengths:**
3. The F-16 Fighting Falcon has been one of the world's most produced fighter aircraft over the past half a century with almost 4500+ aircrafts produced and operated by 25+ nations.
4. F-16's unique positioning as the most cost effective lightweight fighter jet, in terms of procurement as well as sustainment, given that the assembly line has amortized itself many times over which enables Lockheed Martin to offer smoking deals to customers.
5. The huge installed base of F-16 program globally gives Lockheed Martin a ready user base seeking for modernizations as well as upgrades which is a huge market in itself. Lockheed Martin had about 325 F-16 upgrade contracts worth $6.5 billion as of 2017.
6. F-16 is a highly manoeuvrable aircraft program, based on USAF Col. John Boyd's Energy Manoeuvrability theory, with its blended wing-body design and second-highest Thrust-to-Weight ratio of 1.05 which comes just below the Eurofighter Typhoon, which has one of the highest thrust-to-weight ratio of 1.15, coupled with canards for extreme manoeuvrability.
7. F-16's unique aerodynamic characteristics with true negative stability during subsonic flight phase, leading to negative static margin and significant tail lift, while during supersonic phase of the flight the centre of pressure shifts & becomes positively stable leading to reduced downforce required to maintain a given AoA and also reduced trim drag with better lift leading to better turn and cruise performance. The F-16 also leverages the vortex lift generated by the strakes, in addition, to the wing lift. During certain phases of the flight, the F-16 generates almost a third of its total lift only & purely from the fuselage while improving actual wing loading by almost 30% from 65 lb/sq. ft. to around 40 lb/sq. ft. This, when combined with a high thrust-to-weight ratio, makes the F-16 an extremely

manoeuvrable & capable fighter jet.

8. The F-16 also provides decent payload capacity of 17,000 lb which almost matches the payload capacity of twin-engine F/A-18E/F program.

9. The F-16 is also one of the fastest fighter jet programs with a high altitude top speed of Mach 2.05 which is just below the leader of the pack, F-15E's Mach 2.5.

10. The overall cost & value proposition from the procurement & TCO perspective is highly favourable for emerging nations which has witnessed a surge in order intake for the F-16 over the recent years with a change in geopolitical environment and the whittling down of traditional world order.

Weaknesses

• The F-16's very limited combat range of 295 Nautical Miles which is the lowest among the fighter jet programs being compared and means frequent aerial refueling which add to the operating costs and limit its ability to operate in specific areas which could be supported by aerial refueling.

• The single engine translates into a single point of failure without power redundancy which could mean a downside for some nations and the preference for twin engine aircrafts.

• The F-16 program orders have so far remained more or less limited to emerging markets and developing nations as the Western World has moved towards 4 plus and fifth generation fighter jet aircraft programs.

• The F-16's design is over half a century old with the aircraft originally designed in the mid-1970s as the world's very first fly-by-wire aircraft program. Lockheed Martin needs to add capabilities incrementally to keep it competitive against the more advanced 4.5 generation aircraft programs.

4. Dassault Rafale

Image credits: Rafale-RIAT 2009

Author: Tim Felce

Source: Wikimedia Commons

Usage: CCA-SA 2.0 Generic

Description:

The Dassault Rafale is a French, multi-role, medium-sized twin-engine fighter aircraft program produced by Dassault Aviation

Primary Role:-

- Air supremacy

- Air-to-Air Interdiction

- Aerial reconnaissance

- Ground support

- In-depth strike

- Anti-ship strike

- Nuclear deterrence

Origin, Timeline & Scope

The program was conceived originally by the French Navy & the French Air Force in the late 1970s as the replacement for their existing fighter aircraft fleets. The plan was to develop the next-generation fighter jet as a multi-nation cooperative effort which led the way to the genesis of Eurofighter Typhoon. However, disagreements over work-share agreements led the French to walk alone on this path which led to Dassault taking the lead contractor role to produce the Rafale. The Rafale entered service in 2001 and is available in 3 variants, namely, single & twin seat variants for land based operations by the Air Force and a Carrier-based variant for naval aviation operations.

Quantities Produced & In-Service Fleet:

The Rafale has primarily been deployed and used by the French Air Force and the French Navy since the production started in 1986. The international exports of Rafale have recently received sales traction with Air Forces of Egypt, India, Greece and Qatar ordering Rafale aircrafts. A total of around 237 Rafales have been produced since the program's EIS in May 2001.

Specifications:

General characteristics[1]

Crew: 1 or 2

Length: 15.27 m (50 ft 1 in)

Wingspan: 10.90 m (35 ft 9 in)

Height: 5.34 m (17 ft 6 in)

Wing area: 45.7 sq.mt.

Empty weight: 10,300 kg (22,708 lb) for B variant (ranges from 21,720 lb to 23,400 lb across variants)

Gross Weight: 15,000 kg./ 33,069 lb

Max takeoff weight: 24,500 kg (54,013 lb)

Powerplant: 2 × SNECMA M88-4e turbofans with 50.04 kN (11,250 lbf) thrust each in dry mode and 17,000 lbf in wet mode

Internal Fuel Capacity: 10,362 lb for Single Seat, 9700 lb for the Two-Seat Variant

Performance

Maximum Speed: Mach 1.8 (1,188 mph) at High-Altitude

Mach 1.1 at Sea Level (1,390 km/h, 860 mph)

Combat Range:

1,000 NM (1,150 miles or 1,850 km) with three drop tanks (5,700 L), two SCALP-EG and two MICA AAMs.

Ferry range: 2,000 NM (2,300 miles or 3,700 km) with 3 drop tanks

Service ceiling: 50,000 ft

G Limits: +9 to -3.6 (+11 MAX)

Rate of Climb: 60,000 ft/min

Thrust/Weight Ratio: 0.988 for Version B, 0.91 based on MTOW

Wing Loading: 67lb/sq.ft.

Approach Speed: Less than 120 Knots

Landing Ground Run: 450 meters without drag-chute

Sources: Dassault Aviation, French Navy, International Directory of Military Aircraft

Key Operators:

French Air Force and French Navy are the Primary Operator of Rafale (180 ordered of the planned 286). Recently, Egypt (54 ordered & 24 delivered), Qatar(36 Ordercd & Options for 36 more), India (36 Ordered & 26 delivered) and Greece (18 Ordered) have ordered the Rafale jets.

Armament & Avionics:

14 Hard points for the Air Force Version with a total payload capacity of 20,900 lb. of External Fuel & Ordnance. The armaments which can be carried by Rafale include:-

Guns

1 X 30mm GIAT 30/M971 Autocannon

Missiles:

Air-to-Air Missiles

- Magic II

- MBDA MICA IR or EM

- MBDA Meteor

Air-to-Surface Missiles:

- MBDA Apache

- MBDA Storm Shadow/SCALP-EG

- AASM-Hammer (SBU-38/54/64)

- AS-30L

- Mark 82

Anti-Ship Missiles

MBDA AM 39-Exocet anti-ship missile

Nuclear Deterrence: ASMP-A Nuclear Missile

Others: Thales Damocles Targeting Pod, Thales AREOS reconnaissance pod, TALIOS multi-function targeting pod, 5 Drop Tanks and Air-to-Air Refueling among Aircrafts

Avionics:

- Thales RBE2-AA AESA radar

- Thales SPECTRA Electronic Warfare system.

- Thales/SAGEM-OSF Optronique Secteur Frontal infra-red search and track (IRST) system

Bombs:

- GBU-12 Paveway II

- GBU-16 Paveway II

- GBU-22 Paveway III

- GBU-24 Paveway III

- GBU-49 Enhanced Paveway II

Program Status:

Currently in production as well as in-service as amongst the world's longest running fighter jet aircraft program

OEM – Dassault Aviation SA

Dassault has been producing the Rafale aircraft program at its production facility at Argenteuil, near Paris.

Key Mission Profiles

- Multi-Role Aircraft Program – Dassault terms it as Omni-Role Fighter

Key Recent Contract Awards

- Dassault has been pushing for international exports of the Rafale program over the recent years (since 2015).

- Qatar placed an order for 24 Rafale jets in 2015 followed by another order for 12 more aircrafts in 2018. It also has options for 36 more aircrafts. Dassault had delivered around 25 aircrafts to Qatar as of February 2020.

- Egypt had placed an order for 54 Rafale jets with 24 aircrafts already delivered.

- 28 more Rafale jets are to be delivered to the French Forces by 2024.

- Indian Air Force ordered 36 Rafale aircrafts in fly away condition in 2016 with all to be delivered by 2021. 26 had been delivered to India by July 2021.

- Greece placed an order for 18 Rafales in 2020 with the first one delivered in July 2021. In a further development, Greece announced order for an additional 6 Rafale aircrafts in early September 2021 taking the total tally of Rafales ordered by Greece so far to 24 now.

Image Credits: USAFE AFAFRICA

Dassault Rafale

Strengths:

• Dassault's significant technological capabilities & experience in the design, development & production of fighter jets dating back to the Mirage aircraft program.

• Strong exports orientation and focus given limited size & scale of European defense spending which has made European OEMs traditionally more oriented towards international exports, unlike their U.S. counterparts. The European A&D OEMs are better positioned with much deeper roots and relationships with key stakeholders & decision-makers across key international markets.

• Rafale , as a typical European system, provides a well-balanced package to the buyer in terms of one of the best & longest unrefueled combat range of all the aircraft programs at 1,000 NM+ which is even better than light jets.

• Rafale has the highest external payload capacity of all aircrafts (at 21,000 lbs) being compared with the exception of F-15E. Rafale has & offers one of the best payload capacity among medium sized fighter jets at 9,500 kg which is almost the same offered by F-15E/EX Strike Eagle/Eagle II platform, which is one of the world's best heavy fighter jets for air superiority and is also termed as the 'flying bomb truck' by the Israeli Air Force.

• Rafale's M88 engine, produced by Snecma, is one of the most reliable & proven military turbofan engine besides the GE's F414 & Pratt & Whitney's F100 engine.

• Rafale carries proven and cutting edge European avionics made by Thales and missiles package from MBDA which further add to the airframe's capability.

• Rafale provides one of the best combat range & payload capacity while offering a decent operating cost per hour (for a medium fighter jet).

Weaknesses:

• Limited production and global in-service fleet of Rafales takes the per unit procurement cost of Rafales on the higher side which has impacted its uptake across exports markets. At around EUR 100 million, the per unit cost of Rafale is much higher than other 4+ generation aircrafts, including the 5th gen F-35, which is available at $80 million per unit. The F-15EX, too, is available at around $86 million per unit.

- The French Air Force and Navy have been the chef patrons of Rafale program with a majority of the units produced going to the French forces. The export base has been relatively very limited and has picked up only over the past half a decade or so. The limited ability of the French Armed forces to sustain the Rafale program makes the Rafale FAL more or less reliant on international exports for survival & sustenance. Thus, there is limited long range visibility & clarity over the sustainability of the Rafale program's production going forward without French State patronage.

- The European defense equipment have traditionally been perceived to be expensive across international markets which limits their competitiveness across price sensitive buyers. The French capital region traditionally has been one of the most expensive defense industrial base in the world and that translates into the price tag of the French & European origin defense equipment on a comparative basis against American equipment backed by the mammoth U.S. defense budget.

- The traditional EUR-USD exchange rate equation has traditionally favored the U.S. as against the EU and has impacted the competitiveness as well as profitability of European exports while competing across international markets.

5. Eurofighter Typhoon

Image Credits: Pexels.com, Image ID: curioso-photography-343648

Description:

The Eurofighter Typhoon is a 4.5 generation, twin-engine Pan-European fighter jet program conceived in the 1980s.

Primary Role

It was conceived originally as an Air-Superiority fighter program with the ground strike mission later added to it, much like the evolutionary path of the F-15E Strike Eagle. It also undertakes Aerial Reconnaissance missions apart from its traditional Air Defense role.

Origin, Timeline & Scope

The program was conceived by the U.K., Germany, Italy, France (later exited) and Spain originally as an Air Superiority Fighter. The program began formally in 1983 under the Future European Fighter Aircraft program as a replacement for the European Tornado fighter jet program developed by the U.K., Italy & Germany collectively. The prototype aircraft flew in 1994 with production contracts awarded in 1998. The program's Entry into Service (EIS) was in August 2003. The aircraft has been developed for air superiority role with high maneuverability for dogfights and air to air combat. The air to strike mission profile was later added to the Eurofighter Typhoon's mission portfolio with the incorporation of relevant missiles & armaments, including, the Brimstone & Storm Shadow missiles.

Quantities Produced & In-Service Fleet: :

The Eurofighter Typhoon has primarily been deployed and used by the Royal Air Force, German Air Force (Luftwaffe) and the Italian Air Force. The international exports of Typhoon have seen limited success with Kuwait, Qatar, Omanand Saudi Arabia. A total of 571 Eurofighter Typhoon aircrafts have been produced & delivered as of October 2020.

BIRDS OF FRAY - TOP GUN: MAVERICK - SPECIAL EDITION

Image Credits: Author - SAC Tim Laurence/MOD

Source: Wikimedia Commons

Terms of Usage: OGL v1.0[1]

Specifications:

General characteristics

- **Crew:** 1 or 2

- **Length:** 15.96 m (52 ft 4 in)

- **Wingspan:** 10.95 m (35 ft 11 in)

- **Height:** 5.28 m (17 ft 4 in)

- **Wing area:** 51.2 sq.mt.

- **Empty weight:** 11,000 kg (24,251 lb)

- **Gross Weight:** 16,000 kg./ 35,274 lb

- **Max takeoff weight:** 23,500 kg (51,809 lb)

- **Powerplant:** 2 × Eurojet afterburning turbofans with 60 kN (13,000 lbf) thrust each in dry mode and 20,000 lbf in wet mode

- **Internal Fuel Capacity:** 11,010 lb

Sources: RAF Typhoon Data, Air Forces Monthly, International Directory of Military Aircraft, Brassey's Modern Fighters

Performance

Maximum Speed: Mach 2 (1,320 mph/2,125 km/h) at 11,000 mt. Altitude

---

Mach 1.25 at Sea Level (1,530 km/h, 950 mph)

Combat Range:

750 NM (863 miles or 1,389 km) for Air Defense role/Ground Attack mission with three drop tanks (1,000 L)

Ferry range: 2,050 NM (2,350 miles or 3,790 km) with 3 drop tanks

Service ceiling: 65,000 ft

G Limits: +9G to -3G

Rate of Climb: 62,000 ft/min

Thrust/Weight Ratio: 1.15 with 100% internal fuel, 1.10 based on MTOW

Wing Loading: 64 lb/sq.ft.

Sources: RAF Typhoon Data, Air Forces Monthly, International Directory of Military Aircraft, Brassey's Modern Fighters

Key Operators:

- Royal Air Force

- German Air Force (Luftwaffe)

- Italian Air Force

- Spanish Air Force

- Austrian Air Force

- Saudi Arabia, Oman, Qatar and Kuwait

Armament & Avionics:

13 Hard points with a total payload capacity of 19,800 lb. of External Fuel & Ordnance. The armaments which can be carried by the Typhoon include:-

Guns

1 X 27mm Mauser BK-27 revolver cannon

Missiles:

Air-to-Air Missiles

- AIM-120 AMRAAM
- MBDA Meteor
- IRIS-T
- AIM-132 ASRAAM
- AIM-9 Sidewinder

Air-to-Surface Missiles:

- AGM-65 Maverick
- AGM-88 HARM
- Brimstone
- Taurus KEPD 350
- Storm Shadow/Scalp EG

Anti-Ship Missiles

Marte ER

Others: Up to 3 Drop Tanks and Conformal Fuel Tanks (Tranche 3 onwards)

Avionics:

- Euroradar CAPTOR radar

- Passive Infra-Red Airborne Tracking Equipment

- Praetorian DASS

- Damocles (targeting pod)

- LITENING III laser targeting pod

- Sniper Advanced Targeting Pod

Bombs:

- Paveway II/III/Enhanced Paveway series of laser-guided bombs (LGBs)

- 500-lb Paveway IV

- Small Diameter Bomb (planned)

- Joint Direct Attack Munition (JDAM) (planned)

- HOPE/HOSBO, in the future

- Spice 250

Program Status:

Currently in production as well as in-service as primarily a true-blue European aircraft with European DNA

OEM – Eurofighter Jagdflugzeug GmbH

The Eurofighter Typhoon project is managed by the Eurofighter Jagdflugzeug GmbH. NATO Eurofighter and Tornado Management Agency manage the program representing the partner nations, namely, U.K., Germany, Italy & Spain.

Key Mission Profiles

- Air Superiority

- Air Reconnaissance

- Ground Strike

- Air Defense & Interdiction

Key Recent Contract Awards

- The Eurofighter Typhoon program has received a much needed boost with the award of a procurement contract by Germany for 38 Typhoons worth EUR 5.4 billion or $6.35 in late 2020 which further extends the Typhoon final assembly line through the mid-2020s.

- The partner nations plan to operate the Typhoons through 2060 with plans for addition of additional capabilities in an incremental manner. The earlier decision for the addition & incorporation of E-Scan AESA Mk1 radar into the program has provided further confidence to operating nations and potential export customers in the program. The program is also going to further develop the E-Scan radar to Mk2 version.

- Eurofighter Typhoon looking actively at winning the Finland's program to replace its ageing legacy F-18 Hornets.

- The Eurofighter Consortium plans to modernize the Manching, Germany based production facility while preparing the final assembly line again & increase workforce by 2023.

- The European partner nations on the Typhoon program will have to really & drastically cut-down & shorten the turn-around cycle for the incremental incorporation of technological upgrades by really fastening the processes. The program management has itself realized it and plans to cut down the turn around time for the delivery of a full enhancement package on existing aircrafts from 2 years at present to a 12-month cycle.

• The industry consortium needs to focus on preparation for the adoption of new software architecture on the Typhoon program for enhanced user-friendliness and broadening of it for incorporation of additional mission profiles.

Eurofighter Typhoon

Strengths:

• The Eurofighter Typhoon undoubtedly is one of the world's most manoeuvrable fighter jet program with its Delta wing, Canards and a top speed of Mach 2 and one of the highest thrust to weight ratio of 1.15 making the Typhoon an excellent dogfighter in aerial combats during within-visual-range engagements even without the availability or use of thrust vectoring capability.

• The Typhoon is backed by a consortium of European nations; rather than a single nation like France in case of Rafale which is more or less pivoted towards the French state for sustenance. The leading EU nations have been fully backing the program which also reduces the program risks for new Typhoon buyers going forward.

• The Typhoon also has one of the highest service ceilings of all aircrafts at around 65,000 feet while carrying a decent payload of around 19,800 lb/9000 kg on its 13 hard points on the airframe.

• The powerplant propelling the Eurofighter Typhoon is also produced by an pan-European industry consortium comprising of industry specialists.

Weaknesses

• Being a true-blue, all-European fighter jet program, the Typhoon has faced the same set of issues as the European Union, especially, painfully slow and bureaucratic decision-making processes over key decisions pertaining to the program. The Franco-German disagreements over work share agreements as witnessed on the latest FCAS program have been the hallmark of almost all, joint European defense programs.

• Lack of clarity and clear visibility over future upgrade pathways over the Typhoon program, especially, when two sixth generation aircraft programs have almost taken-off across continental Europe. The process of last approved upgrade of the incorporation of the E-Scan AESA radar on the Typhoon, too, was slow & uncertain.

• At around EUR 120 million per unit for export customers, the Typhoon is one of the world's most expensive 4.5 generation aircraft program.

• The Typhoon does not have thrust vectoring capability (TVC), unlike the Russian Su-30/Su-35 Flanker family, in a head to head comparison, with the latter having TVC, Supercruise and Supermaneuverability while being available at a price point of around $86 million per unit.

• The European partner nations on the Typhoon program will have to really & drastically cut-down & shorten the turn-around cycle for the incremental incorporation of technological upgrades by really fastening the processes. The program management has itself realized it and plans to cut down the turn around time for the delivery of a full enhancement package on existing aircrafts from 2 years at present to a 12-month cycle.

• The industry consortium needs to focus on preparation for the adoption of new software architecture on the Typhoon program for enhanced user-friendliness and broadening of it for incorporation of additional mission profiles.

6. Sukhoi Su-35 Super Flanker / Flanker-E (NATO)

Image Credits: Author – Dmitry Terekhov, Source: Wikimedia Commons, Usage terms: CC BY-SA 2.0

Description:

Sukhoi's Su-35 is primarily a single seat, upgraded variant of the traditional Su-27 Flanker aircraft program. The Su-35 features a redesigned cockpit with much improved human-machine interface, weapons control system, a PESA radar and thrust vectoring engine nozzles. The upgraded variant was re-designated as Su-35 by Sukhoi for better international export prospects.

Primary Role

It has been developed as an all-weather, multi-role fighter jet aircraft primarily for air superiority and ground attack roles for international exports.

Origin, Timeline & Scope

The Su-35 program had its origin first in the 1980s as the Su-27M featuring upgrades over the traditional Su-27 Flanker. The second attempt at the creation of the Su-35, the Su-35S, dates back to 2003 under Sukhoi's T-10BM program aimed at creating an interim aircraft program featuring multiple upgrades over the Su-27 geared towards bridging & narrowing the qualitative gap between Russian fighter jets and the Western 4th generation fighter programs till the 5th generation Su-57 entered production & service.

First Su-35S prototype was completed in mid-2007, had its maiden flight in Feb 2008. The Su-35S was showcased at the Paris Air Show 2013. The series production of the Su-35S began in 2010. Russian Air Force became the launch customer of the Su-35S by placing an order for 48 aircrafts at the 2009 MAKS Air Show. The Su-35S achieved Full Operational Capability (FOC) in late 2018.

Quantities Produced & In-Service Fleet:

The first 6 production aircrafts were delivered by Sukhoi to Russian defence ministry by the end of 2012. Around 12 Su-35S aircrafts were delivered in 2013 followed by another dozen in 2014.

United Aircraft Corporation (UAC) had produced around 100 aircrafts by the end of 2018. The deliveries of the Su-35S to Russian Air Force were completed by the end of 2020. Egyptian Air Force placed order for 17 Su-35 aircrafts and has already received them. China also has been a Su-35 operator having placed order for 24 aircrafts which were delivered to it in 2019.

Image Credits: Author - Dmitry Terekhov, Source – Wikimedia Commons, Usage: CC BY-SA 2.0

Specifications:

General characteristics

Crew: 1

Length: 21.9 m (71 ft 10 inches)

Wingspan: 15.3 m (50 ft 2 inches)

Height: 5.9 m (19 ft 4 inches)

Wing area: 62 sq.mt. (670 sq. ft.)

Empty weight: 19,000 kg (41,888 lb)

Max takeoff weight: 34,500 kg (76,059 lb)

Powerplant: 2 × Saturn NPO AL-41F1S afterburning turbofans with 86.3 kN (19,400 lbf) thrust in Dry mode

137.3 kN (30,900 lbf) with afterburner and 142.2 kN (32,000 lbf) in emergency power.

Internal Fuel Capacity: Around 11,500 kg (25,400 lb)

Payload: 8,000 kg (17,630 lb)

Sources: KnAAPO, Jane's All the World's Aircraft 2013

<u>Performance</u>

Maximum Speed: Mach 2.25 at altitude (2,400 km/hr or 1,500 mph)

Combat Range: 860NM on Internal Fuel (990 miles or 1,600 km)

Range: Around 1,900NM (2,200 miles or 3,600 km) on Internal Fuel

Ferry Range: 2,400 NM with 2 external drop tanks

Thrust/Weight Ratio: 0.92 with full internal fuel/1.13 with 50% fuel

Wing Loading: 408 kg/sq.mt. with 50% fuel, 500.8 kg/sq.mt. with full internal fuel

G-Limits: +9

Service Ceiling: 18,000 m/59,000 feet

Sources: KnAAPO, Jane's All the World's Aircraft 2013

Key Operators:

- Russian Air Force

- PLAAF (China)

- Egyptian Air Force

Variants

The Su-35 is already being produced in the following variants:-

- **Su-27M/Su-35:** Single Seat Fighter Aircraft

- **Su-37:** The technology demonstrator with enhanced pilot control. It featured a digital fly-by-wire control system, glass cockpit, N011M radar and AL-31FP engines with thrust vectoring.

- **Su-35UB:** Twin-seat trainer variant designed & built by KnAAPO. The variant featured canards and tall vertical tails featured on Su-27M

- **Su-35S:** The Su-35 variant developed & produced for the Russian Air Force

Armament & Avionics:

12 Hardpoints, including, two wingtip rails, and 10 under-wing & under-fuselage stations with capability to carry external drop tanks & weapons. Total payload capacity of 8,000 kg or 17,630 lb.

The armaments which can be carried by the Su-35 include:-

1 X 30mm Gryazev-Shipunov GSh-30-1 Autocannon

Air-to-air missiles:

- 8 X R-27ER/ET

- R-40

- R-60

- 6 X R-73E

- 12 X R-77M/P/T

- 6 X R-74

Air-to-surface missiles:

- Kh-25ML

- 6 X Kh-29L/TE

- 3 X 3M-14AE

Anti-Radiation Missile

- Kh-25MP
- 6 X Kh-31P/PD
- 5 X Kh-58UShE

Anti-Ship Missiles:

- 3 X 3M-54AE1

- 6 X Kh-31A/AD

- Kh-35U

- 5 X Kh-59MK

- 1 X Yakhont

- Oniks Anti-Shi Cruise Missile

Bombs

- 8 × KAB-500KR TV-guided bombs

- 8 × KAB-500L laser-guided bombs

- 8 × KAB-500OD guided bombs

- 8 × KAB-500S-E satellite-guided bombs

- 3 × KAB-1500KR TV-guided bombs

- 3 × KAB-1500L laser-guided bombs

- GBU-500 laser-guided bomb

- GBU-500T TV-guided bomb

- GBU-1000 laser-guided bomb

- GBU-1000T TV-guided bomb

Avionics:

- N035 Irbis-E Passive Electronically Scanned Array Radar (PESA) equipped with Synthetic Aperture Mode which is likely to be replaced by an AESA radar from the Su-57 program by 2027.

- OLS-35 Infra-Red Search and Track System and Optoelectronic Targeting System which enables the Su-35 to carry advanced weapon systems like the Kh-32 Anti-Ship cruise missile.

- L175M Khibiny-M Electronic Countermeasure System

Program Status:

The Su-35 program is under series production with deliveries of aircrafts to Russian Air Force and PLAAF already completed while deliveries to the Egyptian Air Force are currently going on with the first five aircrafts delivered by February 2021.

OEM – Sukhoi Design Bureau under United Aircraft Corporation

Su-35 was the second aircraft program to be produced by the United Aircraft Corporation after MiG-35. The Su-35 is being produced by the Sukhoi Design Bureau at the Komsomolsk-on-Amur Aircraft Plant.

Key Mission Profiles:

- Air Superiority

• Aerial Combat

• Ground Strike Missions

Key Recent Contract Awards, Related Developments & the Way Ahead

• Egypt placed an order for around two-dozen Su-35 aircrafts from Russia in late 2018, which was reported in March 2019, in a $2 billion deal. The production of the Su-35s for the Egyptian Air Force began in May 2020 with first 5 aircrafts delivered to Egypt by February 2021. Russia plans to deliver another 8 aircrafts to Egypt in 2021

• The Russian Aerospace Forces are beginning to ramp up procurement of the advanced fifth generation Su-57 platform over medium term which is likely to be followed by the scratch-up, sixth generation MiG-41. Sukhoi, thus, in the wake of things is planning to merge the assembly lines of the Su-35 and Su-30 programs from 2027 onwards with Sukhoi looking at harnessing economies of scale with both Su-35 and the Su-30 being very similar designs derived from the original Su-27 Flanker. Sukhoi is likely to merge both the Su-30 & Su-35 programs into a single fighter program by 2027.

• Russian Air Force has announced that it plans to upgrade its in-service fleet of Su-30SMs to the SM2 or Super Sukhoi standard going forward. The Russian Defense Ministry gave a green light to the program in August 2021 and as per plans the upgrade & modernization of the first set of Su-30SMs will commence in December 2021 with the first set of upgraded Su-30SM2s likely to be available for testing & certification by the end of 2022.

Su-35 Super Flanker/Flanker – E

Strengths:

• The Su-35 is truly a 4+++ generation fighter jet aircraft program blending the proven Su-27 Flanker family's proven airframe with increased usage of composites, powerful engines, limited stealth coatings for a reduced RCS and modern avionics for a much superior and highly capable 4.5 generation multi-role combat aircraft platform with a much enhanced endurance.

• The Su-35's powerful Saturn AL-41F1S engines enable the aircraft to undertake super-cruise without using the afterburners which is a capability more or less limited to the fifth generation fighters only .

• The Su-35, with its price point of around $86 million, provides a heavy fighter jet with a proven airframe and much superior 4+++ generation capability at a much lower overall price point than the European origin 4.5 generation medium fighter jet programs, like Dassault Rafale and the Eurofighter Typhoon, both of which have price tags ranging between $100 million to $120 million per unit without the ability to supercruise.or having supermaneuverability.

• The Su-35's thrust vectoring engine nozzles, with their rotational axes canted at an angle and integrated flight & propulsion control systems, enable the Su-35 to have supermaneuverability, i.e., the ability to undertake post stall, low-speed manoeuvres.

• The development of the Su-35 bridges the capability gaps between the fourth and fifth generation fighter jets marginally with a reduced radar cross section, advanced avionics and powerful engines providing super-cruise and super-manoeuvrability, much like what Boeing has tried to do with the creation of the F-15SE Silent Eagle.

• The Supercruise ability provides the Su-35 with the ability to engage its opponents at much greater speed and from a much higher altitude which effectively increases the range of its long-range missiles almost 30% to 40% comparatively.

Weaknesses:

• The Su-35 is a 4+++ generation program with marginally reduced RCS, externally mounted weapons and has an onboard PESA radar which limits its capabilities against a latest 5th generation fighter jet aircraft program like the Lockheed Martin's F-35 or China's J-20. The incorporation of the PESA radar specifically limits its situational awareness & capabilities against its contemporary 4.5 generation fighter jet programs from the West, like Dassault Rafale or the Eurofighter Typhoon.

• The Su-35 has a relatively high price tag, unlike the typical Russian-origin fighter jets, at almost $86 million which makes it almost comparable to the price zone of fifth generation fighter jet programs, like the Su-57, which is available at between $80-$100 million per unit and the F-35 which is now available at $80 million per unit. The high cost of the Su-35, thus, is its biggest downside from a competitiveness perspective. The lifespan of the Su-35, thus, was limited from the outset contingent on the pace of development & production of 5th generation fighter jets.

• The primary role of the Su-35, with its super-maneuverability & super-cruise capabilities, is more or less limited to the Air-to-Air combat role despite the presence of onboard avionics enabling it to undertake ground strike missions and qualifying it for a multi-role combat program.

• The uncertainty over the future & lifespan of the Su-35 program is more or less likely to be dependent on international exports given that the Russian Air Force is moving towards procurement & induction of the 5th generation Su-57 aircraft program which is likely to more or less eclipse the Su-35 program effectively, especially, given the limited export orders Sukhoi so far has on its order book for it.

7. Lockheed Martin F-35 Lightning II Joint Strike Fighter (JSF)

Image Credits: Pixabay.com

Description:

The F-35 Lightning II is a single engine 5th generation stealth fighter jet aircraft program developed & produced by Lockheed Martin Corporation. It has been pursued & developed as a multi-nation program comprising multiple nations as program partners. The F-35 program will be the world's largest & longest running fighter jet aircraft program at over $1 trillion.

Primary Role

It has been developed primarily as a single seat, single engine, all-weather multirole, stealth, long-range fighter jet aircraft program for air superiority and strike missions. Additionally, the F-35has serious capabilities for Electronic Warfare, ISR with its cutting-edge sensors, Link 16 and Sensor Fusion capabilities.

Origin, Timeline & Scope

The F-35 had its genesis in the Joint Strike Fighter (JSF) program, an amalgamation of a number of new aircraft development programs, initiated in the mid-1990s to replace a number of U.S. multi-role strike aircrafts, especially, F-16, F/A-18E/F Super Hornet, A-10 & the F-117, with two contenders, namely, Lockheed Martin's X-35 and Boeing's X-32, being the final down selects for the JSF program. The X-35 was announced as the winner of the program in 2001 which is being pursued as a multi-nation joint program, led by the U.S., which plans to procure around 2,456 F-35s through 2044.

Quantities Produced & In-Service Fleet:

• A total of 690+ F-35 Lightning II JSF aircrafts have already been produced so far by Lockheed Martin as of September 01, 2021.

• Primary users of the F-35 Program have been:-

• U.S. Air Force

• U.S. Navy

• U.S. Marine Corps

• Royal Air Force

• Israeli Air Force

• Royal Australian Air Force

Specifications:

General characteristics (For F-35A Variant)

- **Crew:** 1

- **Length:** 15.7 m (51.4 ft)

- **Wingspan:** 11 m (35 ft)

- **Height:** 4.4 m (14.4 ft)

- **Wing area:** 43 sq. mt. (460 sq. ft.)

- **Empty weight:** 13,290 kg (29,300 lb)

- **Gross Weight:** 22,471 kg (49,540 lb)

- **Max takeoff weight:** 31,751 kg (70,000 lb)

- **Powerplant:** 1 × F135-PW-100 afterburning turbofan with 120 kN (28,000 lbf) thrust in dry mode and 43,000 lbf (190 kN) in wet mode.

- **Internal Fuel Capacity:** 8,278 kg/18,250 lb

- **Payload:** 8,200 kg (18,000 lb)

Sources: Lockheed Martin F35 Program Specifications, International Directory of Military Aircraft, FY2019 Select Acquisition Report (SAR), Director of Operational Test & Evaluation

Variants Developed

- **F-35A** – The Air Force variant equipped with CTOL (Conventional Take-Off and Landing) for land based operations

- **F-35B** – The STOVL (Short Take-Off & Vertical Landing) Variant for Land & Carrier based Ops

- **F-35C** – The CATOBAR (Catapult Assisted Take-Off But Arrested Recovery) equipped Naval variant for Carrier based Operations.

Performance

Maximum Speed: Mach 1.6

Combat Range:

669 NM (770 miles or 1,239 km) on Internal Fuel

760 NM (870 miles or 1,410 km) interdiction mission on internal fuel for internal air-to-air configuration

Range: 1,500 NM (1,700 miles or 2,800 km)

Service ceiling: 50,000 ft

G Limits: +9

Thrust/Weight Ratio: 0.87 at Gross Weight and 1.07 at Loaded Weight with 50% internal fuel

Wing Loading: 107.7 lb/sq.ft. or 526 kg/sq.mt. at gross weight

Sources: Lockheed Martin F35 Program Specifications, International Directory of Military Aircraft, FY2019 Select Acquisition Report (SAR), Director of Operational Test & Evaluation

Key Operators:

- U.S. Air Force, U.S. Navy and the U.S. Marine Corps (Around 2,500 Aircrafts planned)

- Royal Air Force & Royal Navy (48 to 80 F-35Bs planned)

- Royal Australian Air Force (72 F-35As ordered)

- Israeli Air Force (75 F-35Is ordered & 75 Planned)

- Italian Air Force (60 F-35As & 15 F-35Bs planned for the IAF and 15 F-35Bs planned for the Italian Navy)

- Republic of South Korea Air Force & Navy (60 F-35As & F-35Bs planned)

- Japan Air Self-Defense Force (105 F-35As & 42 F-35Bs)

- United Arab Emirates Air Force (50 F-35As planned)

- Air Forces of Norway, Poland, Netherlands, Denmark & Belgium

Armament & Avionics:

4 Internal Stations and 6 External Stations on Wings with a total payload capacity of 5,700 pounds (2,600 kg.) internal, 15,000 pounds (6,800 kg.) external and 18,000 pounds (8,200 kg.) total weapons payload capacity. The armaments which can be carried include:-

Guns

1 X 25mm GAU-22/A 4-barrel rotary cannon

Missiles:

- **Air-to-Air**

- AIM-120 AMRAAM

- AIM-9X Sidewinder

- AIM-132 ASRAAM

- MBDA Meteor (Block 4)

- **Air-to-Surface**

- AGM-88G AARGM-ER (Block 4)

- AGM-158 JASSM

- SPEAR 3 (Block 4)

- Joint Air-to-Ground Missile (JAGM)

- Joint Strike Missile (JSM, planned)

- SOM

- **Anti-Ship Missile**: AGM-158C LRASM

Avionics:

AN/APG-81 AESA radar

AN/AAQ-40 E/O Targeting System (EOTS)

AN/AAQ-37 Distributed Aperture System (DAS) missile warning system

AN/ASQ-239 Barracuda electronic warfare system

AN/ASQ-242 CNI suite, which includes

- Harris Corporation Multifunction Advanced Data Link (MADL) communication system

- Link 16 data link

- SINCGARS

- An IFF interrogator and transponder

- HAVE QUICK

- AM, VHF, UHF AM, and UHF FM Radio

- GUARD survival radio

- A radar altimeter

- An instrument landing system

- A TACAN system

- Instrument carrier landing system

- A JPALS

- TADIL-J JVMF/VMF

Bombs:

- Joint Direct Attack Munition (JDAM) series

- Paveway series laser-guided bombs

- AGM-154 JSOW

- Mk.20 Rockeye II cluster bomb

- Mk 77 incendiary bombs

- Wind Corrected Munitions Dispenser (WCMD) capable

- B61 mod 12 nuclear bombs

Program Status:

Currently in production as well as in-service

OEM – Lockheed Martin Corporation

The F-35 program is manufactured by Lockheed Martin at its Fort Worth, Texas based production facility.

Key Mission Profiles:

Multi-Role Program

- Aerial Combat

- Air Reconnaissance

- Ground Strike

Key Recent Contract Awards, Upcoming Milestones & Latest Developments

- International exports have been a key focus area for the F-35 program given the program's inception being pivoted on international cooperation as a multi-nation cooperative program.

- The F-35 has achieved significant success in taking on the 4/4.5 generation aircraft programs effectively as the air forces globally seek replacement for their legacy, in-service 4/4.5 generation aircraft programs.

- Switzerland's latest decision to procure around 36 F-35As for its Air Force worth $6.5 billion has come as a shot in the arm for Lockheed Martin and the F-35 program following the selection of F-35 by a number of EU nations already, including, Denmark, Norway, the Netherlands and Poland. Further, a bevy of nations, including, Canada & Finland are yet to decide regarding their replacement fighters for the existing F-18 Hornet legacy fleets and the F-35 is one of the key contenders in these contests.

- Earlier in January 2021, UAE also joined the F-35 franchise by signing an agreement with the U.S. administration for the procurement of 50 F-35 aircrafts which are scheduled to be delivered to UAE by 2027.

Going forward, the Department of Defense plans to sign the next multi-year procurement agreement, for Lot 15-17, for the F-35s following the last multi-year contract signed in October 2019 for the Lot 12-14 for 478 F-35 aircrafts worth $34 billion. The Lot 12-14 agreement included 291 F-35 aircrafts for the U.S. Military services and 127 aircrafts for international customers. Also, the Lot 12-14 agreement achieved a major program goal of getting the per unit price or URF cost to $80 million per aircraft marking a 12.8% drop in URF cost per unit from the previous Lot 9-11 with LMT's Fort Worth based assembly line peaking in capacity as well as efficiency which is likely to limit the scope for unit price reduction in the future procurement Lots.

Additionally, a majority of the cost reduction came from Lockheed Martin with costs over the F-135 engine reducing only by 3.5% over previous lot. Also, the Lot 15-17 is likely to include a much larger number of F-35 aircrafts. The negotiations for the Lot 15-17 are likely to be conducted in FY2021 with the decision for Lot-15 likely to be completed in FY2021 itself while the Lot-16 & Lot-17 procurement agreements are likely to be signed in FY22 and FY-23 respectively. The F-35 program also is yet to achieve the full-rate production or Milestone C which will enable the DoD to buy the F-35s in multi-year, block buys. The full-rate production on the F-35 program was postponed once again in December 2020 & then in March 2021, as expected, owing to lack of full integration with the DoD's Joint Simulation Environment.

F-35 Lightning II JSF

Strengths

• The F-35 program is truly one of the most capable & affordable 5th generation, stealth fighter jet aircraft program in production globally backed by Lockheed Martin's domain capabilities in fighter jet development and the U.S. DoD's procurement commitment of up to 2,500 aircrafts by 2044 with the F-35 program likely to remain in-service at least through 2070.

• The F-35 has been equipped with some of the best sensors providing significant situational awareness with matchless sensor fusion & Link 16 capabilities coupled with its unmatched, high survivability emanating from its stealth design making it an unbeatable, overall package. Switzerland's latest decision to procure F-35 was largely driven by the combination of program's situational awareness, high survivability and lower TCO calculation which pegged the procurement of 36 F-35As to cost $2.6 billion less than 4/4.5 gen programs on a comparative basis.

• The F-35's stealth design, highly optimized radar cross section and internal weapons bay enable & render the F-35 as highly suitable to carry out penetration of sophisticated enemy air defenses and take-on A2AD threats effectively and open the gateway for the following waves of 4/4.5 generation fighter aircrafts.

• The procurement per unit cost (on a URF per unit basis) of the F-35 program for the Block 12-14 has reduced to under $80 million with the assembly line reaching peak capacity & efficiency. This reduction in list price has made the F-35 program highly competitive against the existing 4/4.5 generation aircraft programs some of which are even higher priced than the F-35 on a per unit basis. For instance, the F-15EX's per unit price in USAF's latest procurement contract has been pegged at $87 million.

Weaknesses

- The sustainment costs of the F-35 program, including operating costs of $44,000 per hour, are relatively much higher than the 4/4.5 generation aircrafts which could make procurement of 4/4.5 gen aircrafts more attractive to potential customers based on operating costs.

- The scope of cost reductions on the F-35 program going forward are likely to be very limited as the production of F-35 has already peaked and the same is likely to limit the relative competitiveness against 4/4.5 gen programs, which have fully amortized production lines, going forward

- The incorporation of U.S. aerospace industry content, especially the engines subjects the F-35 aircraft procurements by international customers to U.S. government clearances while seeking international export opportunities.

- The F-35 final assembly line faced disruptions in 2021 due to COVID-19 outbreak with Lockheed Martin delivering 20 less F-35 aircraft units to customers than planned which will have to be made up for by 2023. The same could cause delivery delays & schedule disruptions going forward following the outbreak of further waves of COVID-19 pandemic.

- Performance & range limitations of the F-35 against conventional 4/4.5 gen fighters owing to stealth design with internal weapons bays than external airframe based hardpoints. LMT has designed the F-35 as a balanced package with significant stealth, sensor fusion & electronic warfare capabilities driven by cutting-edge sensors offset by limited range & payload capacity.

8. Lockheed Martin F-22A Raptor

Image Credits: Pixabay.com

Description:

The F-22A Raptor is a twin engine 5th generation stealth fighter jet aircraft program developed & produced by Lockheed Martin Corporation. The Raptor is the world's best air dominance fighter jet combining stealth, agility, high situational awareness and effective strike capabilities.

Primary Role

It has been developed primarily as a single seat, twin engine, all-weather stealth, tactical fighter jet aircraft program for unmatched air dominance along-with ground strike, electronic warfare and signals intelligence.

Origin, Timeline & Scope

The F-22 program originated from the YF-22 which had its genesis in the Advanced Tactical Fighter (ATF) program, which was touted as a replacement for the USAF's F-15E Strike Eagle and the F-16 Fighting Falcon programs and was initiated in the mid-1980s. Stealth & Supercruise were the most desired attributes of the ATF program. Lockheed Martin & Northrop Grumman were the final two down-selects for the ATF program which produced YF-22 & YF-23 prototypes for the program. Lockheed Martin's YF-22 powered by Pratt & Whitney's YF-119 engine won the ATF program in 1991. The F-22 had its maiden flight in 1997, achieved IOC (Initial Operational Capability) in 2005 & FOC in 2007 with a total of 187 aircrafts produced through 2011.

Quantities Produced & In-Service Fleet:

- A total of 187 F-22A Raptor aircrafts were produced for the USAF excluding 8 test aircrafts.The sole user of the F-22 Program has been the U.S. Air Force

The advanced nature of avionics and other electronic equipment incorporated by the F-22 have ensured that it remains solely in use exclusively by the USAF with the imposition of international exports ban on the F-22 program by the U.S. government.

Specifications:

<u>**General characteristics (For the F-22A)**</u>

Crew: 1

Length: 18.92 m (62.1 ft)

Wingspan: 13.56 m (44 ft 6 inches)

Height: 5.08 m (16 ft 8 inches)

Wing area: 78.04 sq.mt. (840 sq. ft.)

Empty weight: 19,700 kg (43,340 lb)

Max takeoff weight: 38,000 kg (83,500 lb)

Powerplant: 2 × PW F119-PW-100 augmented turbofans with 116 kN (26,000 lbf) thrust in dry mode and 35,000 lbf (156 kN) in wet mode.

Internal Fuel Capacity: 8,200 kg/18,000 lb

Payload: 9,000+ kg (20,000 lb)

<u>Performance</u>

Maximum Speed: Mach 2.25 at altitude

Mach 1.21 at Sea Level

Mach 1.82 supercruise at altitude

Combat Range:

460 NM (530 miles or 850 km) on Internal Fuel with 100 NM in Supercruise mode, 590 NM (679 miles or 1,093 km) clean subsonic

Range: 1,600 NM (1,800 miles or 3,000 km) or more

Service ceiling: 65,000 ft

G Limits: +9/-3

Thrust/Weight Ratio: 1.08 at Gross Weight and 1.25 at Loaded Weight with 50% internal fuel

Wing Loading: 77.2 lb/sq.ft. or 377 kg/sq.mt. at gross weight

Sources: Lockheed Martin F22 Program Specifications, USAF, Aviation Week, Air Forces Monthly and Journal of Electronic Defense.

Key Operators:

U.S. Air Force has been the sole operator of the F-22A with a ban on international exports. Also, the Radar Cross Section of the F-22 is 0.0001 sq.mt. as per Lockheed Martin Corporation which is equivalent of a steel marble.

Variants

YF-22A: Pre-production technology demonstrator for the ATF program

F-22A: Single-seat production version

F-22B: A planned 2-Seat variant which was cancelled in 1996

F-22 Naval Variant: A naval, carrier-borne variant of the F-22 featuring variable-sweep wings was proposed for the U.S. Navy which was ultimately shelved in 1991 over costs

Avionics:

- AN/APG-77 or AN/APG-77(V)1 Radar

- AN/AAR-56 Missile Launch Detector (MLD) later upgraded to Information & Electronic Warfare System developed by BAE Systems

- AN/ALR-94 radar warning receiver (RWR) with a 250 nautical miles (460 km) or more detection range

- Integrated TRW CNI Avionics, including, a Intra-Flight Datalink, Joint Tactical Information Distribution System (JTIDS) link and an Identification Friend or Foe (IFF) System.

- MJU-39/40 flares for protection against IR missiles

- Northrop Grumman built LTN-100G Laser Gyroscope Inertial Reference, GPS and a Microwave Landing System

Program Status:

Currently Out of production with the F-22 program production wound up by DoD in 2009 with last F-22 delivered to the USAF in 2012. The F-22 has been in-service with the USAF.

OEM – Lockheed Martin Corporation

The F-22 was manufactured & serial produced by Lockheed Martin Corporation till 2012.

Key Mission Profiles:

- Air Dominance

- Aerial Combat

- ISR & Electronic Warfare

- Ground Strike

Key Recent Contract Awards & Related Developments

- The F-22 program's assembly line was closed by Lockheed Martin Corporation in 2012 following the winding up of the program by DoD in 2011.

- In December 2019, Lockheed Martin Corporation was awarded a 5-year sustainment support contract by the DoD for $7 billion, under the exercise of an option on an existing agreement based on the Performance-based Logistics (PBL) model. The latest agreement extends Lockheed Martin's existing service contract on the F-22 program through 2032.

F-22A Raptor

Strengths:

- The F-22 program is the world's best tactical dominance & air superiority aircraft providing an unmatched combination of stealth, speed, manoeuvrability, survivability, strike ability and overall performance.

- The F-22 somehow combines the electronic sensors & other cutting-edge capabilities of the F-35 with the tremendous dogfight & air dominance capabilities of the F-15E providing an unbeatable package for the air force using the platform. Top speed of Mach 2.5 with a thrust-to-weight ratio of 1.08 (even higher than F-15E's 0.93) with a MTOW of 83,000 lb and twin-engines generating a huge thrust output of 26,000 lbf each is tough to match for any competing combat aircraft in the world. The F-15E, comparatively, has a MTOW of 81,000 lb with two PW engines producing 17,800 lbf thrust each.

• Highly capable of undertaking enemy airspace penetration, deep strike missions and knocking down A2AD threats, electronic warfare, ground strike missions and aerial combat simultaneously with a high degree of stealth & survivability emanating from very low radar cross section, proprietary stealth coatings and internal weapons bay.

• The F-22A Raptor forms the core of USAF's overall force structure & fighting capabilities as the most modern & cutting edge, 5th generation multi-role aircraft with tremendous capabilities with around 187 aircrafts in active service. The F-22 retains a hallowed status among USAF pilots who have flown it and is likely to stay that way going forward with the F-22 likely to remain in service through 2050.

Weaknesses:

• The procurement, operating & sustainment costs of the F-22 program, including, list price of $143 million (as per USAF) and operating costs of around $70,000 per hour, are astronomical compared to any 4/4.5 generation aircraft or even the F-35 Lightning II JSF which made the USAF curtail the number of F-22 fighter jets procured to just 187 as against 700 aircrafts originally planned. The huge costs of the F-22 program paved the way for the creation of the F-35 program at the turn of the century as a much more affordable program with a list price of under $100 million and reduced operating costs of around $44,000 per hour in 2021

• The restrictions placed by the DoD on international exports of the F-22 have ensured that the USAF is the sole operator of the F-22 program globally which led to a limited and one of the shortest production runs for a fighter jet aircraft program purely for the USAF with the assembly line getting closed in less than a decade and a half of commencing production. The overall user base of the F-22 thus is very limited comparatively comprising only of the USAF.

• The F-22 is actually the holy grail of fighter jets with its seamless & unmatched capabilities as a 5th generation fighter jet globally but is an out-of-production aircraft program.

• Limited range is another limitation of the platform with a combat range of 460NM. The F-22, thus, essentially is a short range air superiority platform.

Maverick's Choice: F-22A Raptor vs. F14D Tomcat & F-15E – Face-Off

Another interesting comparison would be between the F-14D Tomcat, the aerial champion of a bygone era in the air superiority & interdiction role and the F-22 Raptor with the Raptor being one of the best, 5th generation stealth fighter jet in the world at the moment which is absolutely unmatched in the air dominance role. Let's compare the F-22A with two absolute icons & legends of the skies, namely, the F-14D Tomcat & the F-15E on specifications and performance specs below:-

F14D v. F-22A Raptor

Parameters	F14D Tomcat	F-22A Raptor	F15E Strike Eagle
Empty & Gross Weight	43,735 lb/ 61,000 lb	43,340 lb/64,840 lb	31,700 lb/58,000 lb
MTOW	74,350 lb	83,500 lb	81,000 lb
Internal Fuel Capacity	16,200 lb	18,000 lb	13,455 lb
Total Engine Thrust (Dry/Wet)	16,610 lbf/ 28,200 lbf	26,000 lbf/35,000 lbf	14,590 lbf/23,770 lbf (For F100-PW-220) 17,800 lbf/29,160 lbf (for F100-PW-229)
Top Speed	Mach 2.34	Mach 2.25	Mach 2.5+
Combat Range/ Range	500 NM/1600 NM	460 NM/1600 NM	687 NM/2,100 NM
Service Ceiling	53,000 ft.	65,000 ft.	60,000 ft.
Rate of Climb	45,000 ft/min	65,000 ft./min	50,000 ft./min
Wingspan	64 ft. 1.5 in	44 ft. 6 in	42 ft. 9.6 in
Wing Loading	96/107.9 lb/sq. ft.	77.2 lb/sq. ft.	73.19 lb/sq. ft.
Thrust to Weight Ratio	0.92	1.08	1.005 (F100-PW-229)
Hard Points	10	4 under Wing & Internal Weapons Bay	11
Payload Capacity	14,500 lb	20,000 lb	23,000 lb

The F-22 Raptor clearly has been designed to dominate the skies in the air-to-air role given its technical specs with its thrust to weight ratio of 1.08 which is highest of the aircrafts trio being compared translating into very high maneuverability and vertical climb rate with the Raptor capable of reaching its service altitude of 65,000 feet in just a minute which is almost 50% faster than the F-14D Tomcat and 30% faster than the F-15E Eagle. In terms of grow weight; the F-22A is almost in the same range as the F-14D & F-15E with the Raptor being the heaviest bird of the trio weighing at 64,800 lb while the F-14D weighs at 61,000 lb followed by the F-15E at 58,000 lb. Structurally, in terms of empty weight, the F-22A and the F-14D are almost identical in empty weight at 43,000+lb while the F-15E is much lighter weighing at 31,700 lb.

In terms of combat range, the F-22A would be the first to need aerial refueling with the shortest combat range at 460NM while the F-15E has the longest combat range at 687 NM while the F-14D is right in the middle at 500NM. The F-15E would be the ideal choice for a ground strike mission in an uncontested airspace with a 15% higher max payload capacity of 23,000 lb than the F-22A while in an A2/AD environment the 5th generation F-22A would be a no brainer with its lowest ever radar cross section giving its formidable stealth and highly advanced avionics.

The Raptor leads the duo of older birds in terms of MTOW as well with a MTOW of 83,000 lb followed by the F-15E at 81,000 lb and the F-14D at 74,350 lb. The ability of the Raptor to be really quick despite being the heaviest bird of the lot emanates from the engine thrust. The F-22A's Pratt & Whitney F119-PW-100 augmented turbofan engine (each of the two) has a massive thrust output of 26,000 lbf in dry mode and 35,000 lbf in wet mode which is almost 60% higher than the F-14D's 16,610 lbf in the dry mode and 25% more than the F-14D's 28,000 lbf in the wet mode coming from its GE F110 engines. The F-15E's uprated F100-PW-229s, too, is also no match for the F119 with the latter's thrust being almost 45% higher in dry mode and around 20% in the wet mode.

The F-22 Raptor, however, like the F-15 Eagle, does not have a naval variant, which though was proposed in 1988, termed the NATF-22 Sea Raptor, and was set to feature variable sweep wings, like the F-14 Tomcat, for carrier based operations under the Navy's Naval Advanced Tactical Fighter (NATF) program. However, the huge cost of development, high maintenance costs of the variable sweep wing system and its likely impact on compromising the jet's radar cross section & stealth capability, apart from the fact that the Navy's F-14s were already faster than the USAF's F-22A and were also cheaper to produce than a potential 5th generation stealth fighter jet, collectively & effectively grounded the NATF-22 program by 1991.

An artist's rendering of the proposed NATF-22. Image Credits: Lockheed Martin Corporation.

Maverick would have surely loved the roar of the F-22A's F119s on take-off power on the ground and the F-22A's superb maneuverability in the air with the best thrust-to-weight ratio of 1.08, highest vertical climb rate albeit with the lowest top speed of the trio at Mach 2.25 as against the F-14D's Mach 2.34 with its swing wing design and the F-15E's Mach 2.5+.

Verdict

Maverick would surely pick the F-22A Raptor as a replacement for his F-14D Tomcat with its unmatched 5th generation stealth capabilities & advanced, cutting edge avionics which would enable him to successfully carry out Suppression of Enemy Air Defense (SEAD) missions, high maneuverability with the best thrust-to-weight ratio of 1.08 along with low wing loading for carrying out close range Aerial Combat Maneuvers (ACM) against enemy jets, high Mach 2+ speed for air dominance & air-to-air interdiction missions along with AIM-120s onboard for Beyond Visual Range (BVR) strikes on enemy jets.

Maverick, thus, would be able to track & engage multiple enemy jets from beyond visual range and take them down while being absolutely invisible to the enemy in the air-to-air missions. In close range combat, the F-22A would dominate the adversary with its speed, capability to pull 9Gs, low wing loading & high thrust-to-weight and thrust vectoring which collectively provide it with very high maneuverability. In ground strike missions as well, Maverick would be able to take down enemy SAMs & Air Defenses without any hassle with its stealth and precision ground strike munitions.

In case, he is deemed too dangerous to fly the 5th generation F-22A Raptor...or on account of its relatively much higher flight cost per hour or while facing a non-peer adversary with limited air defense, the F-15E Strike Eagle would be his obvious second choice...However, he would not able to buzz the tower and land any of these two birds on to the Navy's carrier decks as they have not been designed for it...But having known Maverick and his exploits for over 3 decades, anything is possible...

9. Chengdu J-20 Mighty Dragon/ Black Eagle

Image Credits: Author - Alert5, Source: Wikimedia Commons, Usage: CC BY-SA 4.0

Description:

The J-20 Mighty Dragon is a single seat, twin-engine 5th generation stealth fighter jet aircraft program developed & produced by China's Chengdu Aerospace Corporation for the People's Liberation Army Air Force (PLAAF).

Primary Role

It has been developed primarily as an all-weather, stealth, multi-role air superiority fighter jet with substantial, precision strike capabilities.

Origin, Timeline & Scope

The J-20 had its genesis in China's J-XX program of the 1990s with Chengdu having proposed a 5th generation stealth fighter jet aircraft program under its Project 718. The J-20 had its maiden flight in January 2011. The J-20 was showcased at the 2016's China International Aviation & Aerospace Exhibition. The J-20 entered service in March 2017, thereby, becoming the world's third, operational 5th generation stealth fighter jet aircraft program after F-22 & F-35. The NATO reporting name of J-20 is Black Eagle.

Quantities Produced & In-Service Fleet:

- A total of around 150 J-20 aircrafts have been produced for the PLAAF since 2009 and as of 2021.

- Primary users of the J-20 Program have been:-

- People's Liberation Army Air Force (PLAAF)

The PLAAF established its first operational combat unit of the J-20 in February 2018.

Specifications:

<u>**General characteristics (For the J-20)**</u>

Crew: 1

Length: 20.3 m (66 ft 7 inches)

Wingspan: 12.88 m (42 ft 3 inches)

Height: 4.45 m (14 ft 7 inches)

Wing area: 73 sq.mt. (790 sq. ft.)

Empty weight: 17,000 kg (37,479 lb)

Max takeoff weight: 37,000 kg (81,571 lb)

Powerplant: 2 × Saturn NPO's AL-31 FM2 afterburning turbofans with 145 kN (32,600 lbf) thrust in wet mode

2 X Shenyang WS-10C afterburning turbofans producing 147 kN (33,000 lbf) with afterburner

Internal Fuel Capacity: 12,000 kg/26,000 lb

Payload: 11,000 kg (24,000 lb)

Performance

Maximum Speed: Mach 2.5 at altitude

Cruise Speed: Mach 2.1+ (2,183 km/hr., 1,356 mph)

Combat Range:

1,100 NM as reported by Tsinghua University

Around 670-700* NM on Internal Fuel

Around 900-1000* NM with Two External Fuel Tanks (2X600 US Gallon Capacity)

Range: 3,000 NM as reported by Tsinghua University

1,600* NM on Internal Fuel

2,100* NM with two external fuel tanks

Service ceiling: 66,000 ft

G Limits: +9/-3

Thrust/Weight Ratio: 0.92 at Gross Weight and 1.12 at Loaded Weight with 50% internal fuel with AL-31FM2/WS-10C engines

Wing Loading: 69 lb/sq.ft. or 340 kg/sq.mt. at gross weight

Sources: Tsinghua University, Own Estimates

Key Operators:

PLAAF has been the sole operator of the J-20 so far

Variants

J-20B: Equipped with Thrust-Vectoring Nozzles powered by the WS-10B-3 engines

A Twin-Engine Variant of J-20 is also under development powered by Shenyang's WS-10C Engines

The WS-15 Powered J-20B to have supercruise capability

Armament & Avionics:

4 External, Under-Wing Pylon Stations with capability to carry external drop tanks.

The armaments which can be carried by J-20 include:-

J-20 has 3 Internal Weapon Bays with the main weapons bay capable of carrying 4 of the following short & long range missiles.-

- PL-9/PL-10 Short Range AAM

- PL-12 C/D Medium Range AAM

- PL-15 BVR Long Range AAM

- PL-21 Long Range AAM

- LS-6 Precision-Guided Bomb

- Anti-Radiation Missile

The Two Smaller, Lateral Bays positioned behind Air Inlets can Carry 1 Short-Range PL-10 AAMs each

The J-20 Features a Glass Cockpit with:-

- One primary LCD Touch Screen

- Three smaller auxiliary displays

- A wide-angle holographic HUD

Avionics:

- Type 1475 (KLJ-5) Active Electronically Scanned Array Radar with 2000-2200 transmit/receive modules

- EOTS-86/89 electro-optical targeting system (EOTS)

- EORD-31 Infrared search and track

- Electro-Optical Distributed Aperture System

Program Status:

The J-20 is in-production and is in active service by the PLAAF with around 150 aircrafts delivered by Chengdu so far as of 2021.

OEM – Chengdu Aerospace Corporation

The J-20 is being manufactured by Chengdu Aerospace Corporation (CAC).

Key Mission Profiles:

- Air Superiority

- Aerial Combat

- ISR & Electronic Warfare

- Ground Strike Missions

Key Recent Contract Awards & Related Developments

• Reports originating Chinese media reported that a new variant of the J-20, termedJ-20B, to be powered by another variant of the Chinese WS-10 C engine, the WS-10B-3, equipped with thrust vectoring nozzles was unveiled in July 2020 and started with serial production in 2019.

• Chinese media also reported in January 2021 that China plans to replace existing Russian-origin AL-31F engines on the J-20, which also power the Su-27 Flanker aircraft family, with the indigenously developed WS-10C and ultimately by Xian made WS-15 engines. The transition to Chinese engines will reduce the J-20 program's and China's dependence on Russian origin engines.

• Chengdu and China are also exploring the potential of a twin-seat variant of the J-20 powered by the WS-10C as reported by the South China Morning Post Infographic. The WS-10C engine variant features improved thrust and stealthier serrated afterburner nozzles.

• The twin-engine variant could be aimed at utilizing the J-20 for additional roles, including,

• Tactical Bombing

• Electronic Warfare

• Carrier Strike Roles, especially, with its frontal stealth and long range capabilities

Also, at this stage, it is unclear whether China will be exporting the J-20 as well or will it keep it only for the PLAAF, just like the U.S. Government ban on the F-22 Raptor's exports. Most likely it will remain exclusive to the Chinese forces as Beijing would not want the workings of its first stealth fighter jet to be decrypted. Further, the Shenyang's FC-31, the smaller and less stealthy counterpart of the J-20 with limited capabilities, has been custom built for exports with private funding, like the F-35, and is likely to be the one to be exported ultimately.

Chengdu J-20

Strengths:

- The J-20 program with its specifications has been well positioned as a 5th generation stealth fighter jet aircraft program for air superiority and ground strike missions and truly is the only match for the F-22 Raptor program globally in terms of capabilities.

- The J-20 has almost 27% more powerful engines and is a long range air dominance fighter jet program, unlike, the F-22 Raptor, which is a short range air dominance program with an operating range of under 500 NM on internal fuel. The J-20 is one of the largest, heaviest & most powerful 5th generation stealth heavy fighter jet aircraft program in the world with one of the longest operating ranges as well.

- The design of the J-20 incorporates canards providing tremendous manoeuvrability, high supersonic performance & improved short-field landing performance to the J-20 for the air dominance role. J-20 is the only 5th generation fighter jet program so far to feature canards. Additionally, the incorporation of thrust vectoring engines on the J-20 (as planned), in combination with the canards, will make the J-20 super-maneuverable for aerial combat and dogfights against other, comparable aircraft programs, like the F-22 or F-35.

- Additionally, the design of the J-20 theoretically features multiple improvements over traditional, western origin stealth fighter jet programs, like, a long, blended fuselage-wing design, which combined with leading edge extensions in a canard layout, provides significantly enhanced lift, which the Chinese designers claim provides 1.2 times the lift of a traditional canard delta and 1.8 times the lift of a pure delta configuration to the J-20. This also allows the J-20 to use smaller wings which also reduce overall drag without compromising maneuverability. The J-20 also has the lowest wing loading factor of all 5th generation stealth fighter jets and is at least 10% lower than the F-22 and Su-57.

• The J-20 provides serious air dominance, aerial combat and strike capabilities to the PLAAF against the west, especially the U.S., in any potential aerial face-off going forward.

Weaknesses

• The J-20 does not have internal cannons for dogfighting and air-to-air engagements for visual engagements and strike ranges. It is dependent completely on missiles for long range stand-off strikes.

• Relatively low thrust-to-weight ratio comparatively of the J-20 as compared to the F-22 reducing its overall effectiveness in terms of maneuverability for the air superiority & dominance role.

• The J-20 currently has been flying with Russian origin Saturn NPO engines with reliability and mission readiness rates being a challenge against western fighters. Chinese aerial capabilities at present are largely dependent on Russian engines & technology.

• The J-20 does not have thrust-vectoring nozzles and super-cruise capability as of now to match the F-22 in an aerial stand-off.

• J-20's going to be way more expensive (given the specifications) than the LMT's F-35 Lightning II JSF or the Russia's latest, single engine Sukhoi Checkmate 5th generation stealth fighter jet shown at the MAKS 2021 held in July 2021.

• The usage of Canards are adding to the structural weight & might be impacting stealth capabilities theoretically and may be redundant following the incorporation of thrust vectoring capability.

• High instability of the J-20's design with a sustained pitch authority at a high angle of attack, which in a typical, conventional airframe would cause stalling. On the J-20, however, it has been offset by and countered with the canards. The F-22, on the contrary, is a relaxed stability design with excellent maneuverability with powerful thrust vectoring engines.

10. Sukhoi Su-57 Felon (NATO)

Image Credits: Author-Anna Zvereva, Source: Wikimedia Commons, Usage: CC BY–SA 2.0

Description:

The Su-57 Felon (NATO) is a single seat, twin-engine 5th generation stealth fighter jet aircraft program developed & produced by Russia's Sukhoi Aircraft Corporation for Russian Aerospace Forces. The Su-57 is the first 5th generation stealth fighter jet aircraft program of the Russian Armed Forces. The Su-57 program 5th generation features, including, Supercruise, Supermaneuverability, stealth and integrated avionics.

Primary Role

It has been developed primarily as an all-weather, stealth, multi-role air superiority fighter jet with precision, ground strike capabilities.

Origin, Timeline & Scope

The Su-57 or the T-50 (Sukhoi's internal program designation) or PAK FA had its genesis in the S-32 program dating back to the 1980s and the MiG 1.44 program, too, going back to the Soviet era. Sukhoi was selected as the winner of the PAK FA program. The maiden flight of the first prototype took place in January 2010 and the T-50 was unveiled at the MAKS Air Show in 2011. The Russian Defense Ministry signed the first procurement contract for 2 Su-57s in August 2018 followed by announcement of 76 more Su-57s to be procured by the Russian Defense Ministry in 2019 with deliveries scheduled through 2028.

Quantities Produced & In-Service Fleet: :

The Russian Defense Ministry has around 78 Su-57 fighter jets on order as of 2021. A total of 12 units of the Su-57 aircraft have been produced by Sukhoi so far, as of 2020, with 10 being the test prototypes and 2 units being serial produced. The Su-57 program is currently in-service with the first serial produced getting delivered to the Russian Air Force in December 2020.

Specifications:

<u>**General characteristics**</u>

Crew: 1

Length: 20.1 m (65 ft 11 inches)

Wingspan: 14.1 m (46 ft 3 inches)

Height: 4.6 m (15 ft 1 inches)

Wing area: 78.8 sq.mt. (848 sq. ft.)

Empty weight: 18,000 kg (39,683 lb)

Max takeoff weight: 35,000 kg (77,162 lb)

Powerplant: 2 × Saturn NPO's AL-41F1 afterburning turbofans with 88.3 kN (19,900 lbf) thrust in wet mode

142.2 kN (32,000 lbf) with afterburner

Internal Fuel Capacity: 10,300 kg/22,700 lb

Payload: Around 7,500 kg to 10,000 kg (16,500 lb to 22,000 lb) Estimated

<u>**Performance**</u>

Maximum Speed: Mach 2 (2,130 km/hr., 1,320 mph) at altitude

Cruise Speed: Mach 1.3 (1,370 km/hr., 860 mph)

Combat Range:

700 to 800 NM on Internal Fuel

Around 900-1000 NM with Two External Fuel Tanks (2X600 US Gallon Capacity)

Range: 1,900 NM (3,500 km) on internal fuel

2429 NM (4,500 km)with two external fuel tanks

Service ceiling: 66,000 ft

G Limits: +9

Thrust/Weight Ratio: 1.02 (1.19 at Normal Take-Off Weight)

Wing Loading: 76 lb/sq.ft. or 371 kg/sq.mt. at Normal Take-Off Weight

Sources: Aviation Week, Aviation News, Key Aero

Key Operators:

Russian Air Force has been the sole operator of the Su-57 so far

Variants

Su-57: First Production Variant with the first aircraft delivered to RAF in December 2020 and orders placed for procurement of 78 aircrafts

Su-57E: Export variant of the Su-57 being promoted as a multi-role aircraft by Rosoboronexport.

Su-57M: The upgraded version of the Su-57 incorporating improved mission systems, enhanced reliability and maintenance, new flight control actuators and the upgraded izdeliye 30 engines. The flight testing for the Su-57M was slated to take place in 2020 with the serial production of the variant scheduled for mid-2020s. A two-seat version for exports is also under development which will feature integration with UAVs & other sub-systems.

Armament & Avionics:

6 External, Under-Wing Pylon Stations with capability to carry external drop tanks & weapons. 6 Internal Weapon Stations. Su-57 has two tandem main internal weapons bays with each measuring 14.4 ft. by 3 ft. and two small triangular section weapon bays that protrude under the fuselage near the wing root. The Su-57 is likely to carry 4 BVR missiles for aerial combat in its two main weapon bays and two short-range missiles in the wing root weapons bays.

The armaments which can be carried by Su-57 include:-

1 X 30mm Gryazev-Shipunov GSh-30-1 autocannon

Air-to-air missiles:

- 8 × R-77M Medium Range Missile with active radar homing

- 2 × Short Range R-73/74M2 Infrared-homing Missile

- R-37M Long Range Hypersonic Air-to-Air Missile

Air-to-surface missiles:

- 4 × Kh-38ME, Kh-59MK2

- KAB-250 (250 Kg) or KAB-500 (500 Kg) & KAB-1500L (1500 kg) precision guided bombs

Anti-ship missiles:

- 2 × Kh-35E/31/UE

Anti-radiation missiles:

- 4 × Kh-58UShK

- Anti-tank "Drill" 500 kg cluster-bomb + active homing

Avionics:

- Sh-121 multifunctional integrated radio electronic system (MIRES)

- Byelka radar (400 km, 60 tracks with 16 targeted)

- N036-1-01: Frontal X-band active electronically scanned array (AESA) radar

- N036B-1-01: Cheek X-band AESA radars for increased angular coverage

- N036L-1-01: Slat L-band arrays for IFF

- L402 Himalayas electronic countermeasure suite

- 101KS Atoll electro-optical targeting system

- 101 KS-O Laser Directional Infrared Counter Measures

- 101KS-V: Infra-Red Search and Track

- 101KS-U: Ultraviolet missile approach warning system

- 101KS-N: advanced stabilized navigation and targeting system

- 101KS-N: optional external Targeting pod

- 101KS-P: thermal imager for low altitude flying and night landing

Program Status:

The Su-57 is in-production and is in active service by the RAF with the first aircraft delivered by Sukhoi in December 2020 and the remaining 77 to be delivered by 2028.

OEM – Sukhoi Design Bureau and Russian Aircraft Corporation

The Su-57 is being manufactured by Sukhoi along with Russian Aircraft Corporation.at the Komsomolsk-on-Amur Aircraft Plant.

Key Mission Profiles:

- Air Superiority

- Aerial Combat

- ISR & Electronic Warfare

- Ground Strike Missions

Key Recent Contract Awards, Related Developments & the Way Ahead

- Sukhoi received the contract for production & delivery of 78 Su-57 aircrafts from the Russian Ministry of Defense in June 2019 which are to be delivered by 2028.

- Sukhoi also is working towards further improving the capabilities of the Su-57, under its Megapolis program, through the incorporation of improved mission systems, enhancements geared towards reliability & maintenance, incorporation of electromechanical drives and the integration of more powerful izdeliye 30 engines producing 30% to 40% more thrust than the current AL-41Fs.

- Sukhoi expects the Su-57 to form the backbone and the core pivot of an entire family of 5th generation fighter jets for the Russian Aerospace Forces going forward by replicating the success of the 4th generation Su-27 Flanker fighter jet aircraft family which received significant success with its variants across domestic as well as international exports markets.

• Sukhoi has also been testing the Su-57 prototypes in teaming arrangements with the Okhotnik UCAV and integration with other sub-systems which will become operational on the two-seat variant with the WSO (Weapon Systems Officer) on the rear seat likely to manage & control these tasks.

Sukhoi Su-57 Felon

Strengths:

• The Su-57 features a blended-wing body design along-with thrust vectoring nozzles which gives the program a much higher lift capacity and excellent manoeuvrability for the air superiority role.

• The Su-57 program has been custom-built by Sukhoi to address the shortcomings & capabilities gaps found on the F-22 Raptor program, which include, the inability to use thrust vectoring to induce roll & yaw movements, insufficient payload capacity and mechanism for stall recovery in scenarios where thrust vectoring fails.

• In terms of operating and combat range, the Su-57 scores over the F-22 with a combat range of around 700-800 NM on internal fuel which is at least 50% more than the F-22 Raptor making it a medium range air dominance & ground strike aircraft unlike F-22's short range.

• The substantial usage of composites on the Su-57 has ensured that the number of parts required have reduced by almost 4 times than the Su-27 Flanker while also reducing the overall weight of the aircraft adding to its range and payload capacity.

• The extensive wing surfaces, huge wing area and leading edge extensions provide the Su-57 a low wing loading of 371 kg/mt. sq. (compare that to F-22 Raptor's 377 kg/mt. sq. and J-20's 340 kg/mt. sq.) which combined with the usage of leading-edge vortex controllers (LEVCONs) to control vortices generated by the leading edge root extensions provide trim along with a sustained, high angle of attack behaviour while equipping the Su-57 with a capable stall recovery mechanism. These features along with an advanced flight control system and thrust vectoring enable the Su-57 to carry out very high angle of attack manoeuvres like Pugachev's Cobra and Bell manoeuvre respectively. The Su-57 can also do flat rotations with little loss of altitude. The wing loading factor of the Su-57, thus, very closely matches the F-22.

• The Su-57 is the first stealth, 5th generation, multi-role fighter jet aircraft program being inducted into the Russian Aerospace Forces and is likely to provide a significant capability upgrade and match against the F-22 and the F-35 aircraft programs while also providing significant international exports opportunities going forward.

Weaknesses

• The Su-57's radar cross section is almost 30 times lower than the 4th generation Su-27 Flanker family, however, at 0.1 to 1 meter square it is still comparatively much higher than the F-22's 0.0001 meter square and the F-35's 0.005 RCS.

• The Su-57 has been criticized for its limited area coverage for stealth coatings mostly emphasized in the frontal area. However, the shaping of the aft fuselage, the seams between parts and rivets are much less optimized for radar stealth compared to the F-22 Raptor. The Indian Air Force, which originally had plans to procure & locally produce the Su-57 as a program partner, opted out of the program owing to the aircraft's limited stealth capabilities.

• Su-57 has a relatively lower thrust to weight ratio as against the F-22 Raptor while also having a relatively lower overall payload capacity than the F-22.

• The Russian jet engines are notorious for their limited reliability and serviceability issues leading to lower overall readiness rates across end users as experienced by India over its Su-30MKI in-service fleet. This could be a serious challenge against western origin fighters.

• The Su-57, in terms of unit cost, at present is at par or slightly more expensive than the F-35 Lightning II JSF, which is being mass produced by the U.S. under an international cooperative set-up, including, traditional NATO allies. The Su-57's international exports potential, thus, is going to be limited with the smaller, single engine Su-75 Checkmate being promoted by Sukhoi & the Russian defense agencies to take the lead in terms of international exports.

Bibliography

1. F-14 Tomcat

1. F-14 Fact Sheet M.A.T.S. https://web.archive.org/web/20061007024558/http://www.anft.net/f-14/f14-serial.htm
2. F-14 Tomcat fighter fact file, U.S. Navy https://web.archive.org/web/20060402215910/https://www.navy.mil/navydata/fact_display.asp?cid=1100&tid=1100&ct=1
3. Warwick., Graham (30 March 1985). "F-14D for digital Grumman's Tomcat is to remain the US Navy's premier fighter". Flight International.

2. F-15E/EX Strike Eagle/ Eagle II

1. Technical Specifications on F-15E by Boeing https://www.boeing.com/defense/f-15/
2. U.S. Air Force Fact Sheet on F-15E Strike Eagle https://www.af.mil/About-Us/Fact-Sheets/Display/Article/104499/f-15e-strike-eagle/

3. Aerofax: McDonnell Douglas F-15 Eagle: Supreme Heavy-Weight Fighter, Midland Publishing, 1998, ISBN-10: 1857800818 ISBN-13: 978-1857800814

3. F/A-18E/F Super Hornet

1. U.S. Navy Fact Sheet on F/A-18E/F Program

https://www.navy.mil/Resources/Fact-Files/Display-FactFiles/Article/2383479/fa-18a-d-hornet-and-fa-18ef-super-hornet-strike-fighter/

1. Boeing F/A-18E/F Technical Specifications

https://www.boeing.com/defense/fa-18-super-hornet/#/technical-specifications[1]

1. Federation of American Scientists – F/A-18 Hornet Overview https://web.archive.org/web/20111026182137/ http://www.fas.org/programs/ssp/man/uswpns/air/fighter/f18.html

4. F-16V Fighting Falcon Block 70

1. U.S. Air Force Fact Sheet on F-16

https://www.af.mil/About-Us/Fact-Sheets/Display/Article/104505/f-16-fighting-falcon/

2. F-16 Aerodynamics - By Joe Bill Dryden, Code One Magazine

https://web.archive.org/web/20080828124207/ http://www.codeonemagazine.com/archives/1986/articles/apr_86/f16_aero/index.html

1. Dassault Rafale

1. Dassault Aviation – Specifications & Performance Data – Rafale

https://www.dassault-aviation.com/en/defense/rafale/specifications-and-performance-data/

1. Characteristics of Rafale M, French Navy

https://www.defense.gouv.fr/marine/equipements/aeronefs/groupe-aerien-embarque/rafale-marine

1. Eurofighter Typhoon

---

1. https://www.boeing.com/defense/fa-18-super-hornet/#_6666cd76f96956469c7bc39d750ee7d9_technical specifications

1. Specifications of Eurofighter Typhoon by Royal Air Force

https://web.archive.org/web/20120810152512/
http://www.raf.mod.uk/equipment/typhooneurofighter.cfm

1. Eurofighter Typhoon – Airpower AT

http://eurofighter.airpower.at/technik-daten.htm

1. FAS Military Analysis Network – Eurofighter Typhoon

https://man.fas.org/dod-101/sys/ac/row/eurofighter.htm

1. Bundeswehr – Eurofighter Typhoon

https://www.bundeswehr.de/de/ausruestung-technik-bundeswehr/
luftsysteme-bundeswehr/eurofighter

1. Aerospaceweb.org – Eurofighter Typhoon

http://www.aerospaceweb.org/aircraft/fighter/typhoon/

1. **Su-35 Super Flanker/ Flanker-E (NATO)**

KnAAPO

https://www.webcitation.org/
6J6cwo1bR?url=http://www.knaapo.ru/media/eng/about/
production/military/su-35/su-35_buklet_eng.pdf

https://web.archive.org/web/20120730185357/
http://www.knaapo.ru/eng/products/su-35/index.wbp

Take Off Magazine, May 2008 Issue

https://web.archive.org/web/20131029195316/
http://en.take-off.ru/pdf_to/to10.pdf

Take Off Magazine, July 2012 Issue

https://web.archive.org/web/20131029193146/
http://en.take-off.ru/pdf_to/to23.pdf

Take Off Magazine, July 2014 Issue

https://web.archive.org/web/20170602213831/
http://en.take-off.ru/pdf_to/to29.pdf

Rosoboronexport

http://roe.ru/eng/catalog/aerospace-systems/fighters/su-35/
presentation#3[2]

1.   F-35 Lightning II JSF

Lockheed Martin Corporation – F-35 Program

https://web.archive.org/web/20110317113904/
http://www.lockheedmartin.com/products/f35/
f-35A-ctol-variant.html

US Air Force

F-35A Variant

https://www.af.mil/About-Us/Fact-Sheets/Display/Article/
478441/f-35a-lightning-ii-conventional-takeoff-and-landing-
variant/

F-35B Variant

---

2. http://roe.ru/eng/catalog/aerospace-systems/fighters/su-35/
presentation#a5c02393e59c943d6a75a9241140faca33

https://web.archive.org/web/20110317114148/
http://www.lockheedmartin.com/products/f35/
f-35b-stovl-variant.html

F-35C Variant

https://web.archive.org/web/20110317173004/
http://www.lockheedmartin.com/products/f35/
f-35c-cv-variant.html

https://web.archive.org/web/20190418155950/https://f35.com/
about/carrytheload/weaponry

https://web.archive.org/web/20190310034412/
https://www.esd.whs.mil/Portals/54/Documents/FOID/
Reading%20Room/Selected_Acquisition_Reports/
18-F-1016_DOC_44_DOD_F-35_SAR_Dec_2017.pdf

1. F-22A Raptor

U.S. Air Force

https://www.af.mil/About-Us/Fact-Sheets/Display/Article/104506/
f-22-raptor/

Lockheed Martin Corporation

https://web.archive.org/web/20120603072326/
http://www.lockheedmartin.com/us/prod ucts/f22/f-22-specifications.html[3]

http://2.bp.blogspot.com/-   irwcM2ov73s/Toru4NRprRI/
AAAAAAAABm4/PBBm_UVa5Ng/s1600/F-22range.jpeg[4]

---

3. https://web.archive.org/web/20120603072326/http://www.lockheedmartin.com/us/prod%09ucts/f22/
f-22-specifications.html

4. http://2.bp.blogspot.com/-%09irwcM2ov73s/Toru4NRprRI/AAAAAAAABm4/PBBm_UVa5Ng/
s1600/F-22range.jpeg

1. J-20 Mighty Dragon

https://guofang.tsinghua.edu.cn/info/1017/1523.htm

http://www.xinhuanet.com/mil/2021-09/28/c_1211385667.htm

https://www.airforce-technology.com/projects/chengdu-j20/

1. Su-57 Felon

https://fighterjetsworld.com/air/sukhoi-su-57-russia-fifth-generation-stealth-fighter- jet/1739/[5]

https://www.militaryfactory.com/aircraft/detail.php?aircraft_id=782

https://aerocorner.com/aircraft/sukhoi-su-57-felon/#aircraft-specifications

https://www.rand.org/blog/2020/08/russias-su-57-heavy-fighter-bomber-is-it-really-a- 5th.html[6]

---

5. https://fighterjetsworld.com/air/sukhoi-su-57-russia-fifth-generation-stealth-fighter-%09jet/1739/

6. https://www.rand.org/blog/2020/08/russias-su-57-heavy-fighter-bomber is it really-a-%095th.html

About the Author

RAJAT NARANG

Rajat Narang is the Co-Founder and Partner of a niche Research Firm pivoted on the Global Aerospace & Defense Industry as part of a career spanning over two decades. He has authored over 1500+ syndicated Research Reports providing strategic analysis to Top Executives across industries, sectors, players & markets with the top users in the Aerospace industry being the OEMs and T1 Suppliers across Commercial, Regional and Business Aviation, including, Airbus, Boeing, Bombardier, Gulfstream, Dassault Aviation, Spirit Aerosystems, GE Aviation, Pratt & Whitney and Safran. Some of the key end users on the defense side have been the U.S. Air Force, Lockheed Martin Corporation, Irish Ministry of Defense, General Dynamics Land Systems, Russian Helicopters and the Korean Aerospace Industries (KAI).

He is the author of multiple books on Commercial and Military aviation. His educational background includes a **Masters in Business Administration - International Business** with Business Strategy as the core pivot followed by a Masters in Political Science with specialization in **International Relations**. Bitten early by the A&D and Strategy bugs while growing up, he has been actively following, tracking & pursuing them for almost 2 decades now.

* Refer Bibliography Point 3 under F-15E at the end

Don't miss out!

Visit the website below and you can sign up to receive emails whenever Rajat Narang publishes a new book. There's no charge and no obligation.

https://books2read.com/r/B-A-KIYU-RCWAC

BOOKS2READ

Connecting independent readers to independent writers.

Also by Rajat Narang

Airbus vs. Boeing: Strategy Wars, Tactical Dogfights, High-G Maneuvers and the Photo Finishes
Birds of Fray - Top Gun: Maverick - Special Edition
Birds of Fray - World's Top 4.5 & 5th Gen Fighter Jet Aircraft Programs

www.ingramcontent.com/pod-product-compliance
Lightning Source LLC
Chambersburg PA
CBHW071523150726